RADIO MEMORY / Brandon LaBelle

RADIO MEMORY / Brandon LaBelle

978-0-9772594-6-5

Errant Bodies Press:
Audio Issues Vol. 4
www.errantbodies.org

Design: fliegende Teilchen, Berlin
Printed: Pinguin Druck, Berlin
Distributed: DAP, www.artbook.com

INTRODUCTION

Brandon LaBelle

I'm always struck by the sounds around us, and in particular, the sudden appearance of songs sounding out from various speakers and systems that come to mark the public environment. *Background noise...* These songs become provocative partners to our own trajectories and experiences, coloring and contouring the ebb and flow of conversation and interactions, daydreams and errands. In this regard the background is not so much a form of wallpaper, however annoying or dull or pleasing, but a dynamic input into personalized and shared narratives. It was with such thoughts in mind that the Radio Memory project developed, initially in relation to the Archipel festival, Geneva, in 2005, which took the theme of "radio" as its focus. In response, the idea of radio memories came to me, and seemed an exciting concept, especially how it weaves together place and audition, knotting experiences with deeper psycho-social input. More so, I was excited to essentially reach out to both friends and strangers and invite their stories into the work – their stories in fact are the work.

The project started by sending out a mass email requesting radio memories – these were generally defined as memories that are partially shaped by being interlocked with a certain song, particularly songs appearing at the moment of the given experience without one's choosing. That is, memories effected by and infused with background noise, as a generative intersection. After receiving many positive responses, I attempted to further my reach by sending out the call to various lists, and asking others to forward it along. A process unfolded that eventually resulted in receiving over 100 memories from around Europe, North and South America, and Japan. As the memories came in, I began to read through and immediately felt a profound sensitivity – suddenly I was holding in my hands some very special stories, each touched by their own sense of emotional retelling, their own sense of personal history, and importantly, imagination. The care with which people wrote their memories, recounting private details, lent a deeper sense of caring for the project in general – in receiving these memories I became a kind of caretaker. The memories are moving, funny, loving, courageous and tender, and, from my perspective, their telling a form of literary prose.

After receiving the memories (most of all the memories appearing here were received between 2004-05) I also began to appreciate how the memories both spoke of highly personal experiences while echoing toward greater, shared coordinates – of heartache, of travels, of sleepless nights… As I prepared the work for exhibition I decided to collect the memories into a form of library, cataloguing them into 10 sections based on loose, poetical themes, such as "Haunted House" or "Pillow Talk" or "Hearing Things". These acted as a way to appreciate the resonances occurring between memories, as well as functioning as potential themes for future radio transmissions: I imagined that in addition to being a caretaker, the memories also provided material for potential broadcasting – to return the memories to the air. This led me to collect each of the songs relating to each of the memories, forming an audible catalogue to the radio memory library.

It was my immediate feeling that the memories and the work should appear in book form, though it has taken 3 years to figure out a way to do so, or maybe the book needed 3 years to take shape, after presenting the project in various guises over the years – the original installation at the Archipel festival, Geneva (2005), as well as installations at the Espace Gantner Gallery in France (2005), at the Radio Revolten festival in Halle, Germany (2006), at Lydgalleriet, Bergen (2008), and at Casa Vecina, Mexico City (2008), as well as radio broadcasts on Resonance FM London and related performances. I'm extremely happy the project continues to have a life, and these memories appear here, finally, as a document of not only my project, but of the movements of sound and music and people's sensitivity for them. In addition, curators Bastien Gallet and Carmen Cebreros Urzaiz, with whom I worked on the project at instances over the years, were kind enough to offer their reflections and meditations on the question of music and memory, and whose texts are included here. As a final addition, a CD of related audio works accompany this publication. This includes a series of works developed recently for the installation in Bergen and which are based on four specific memories from the library. Using these as a generative base, I wanted to extend their stories into dramatic scripts, giving narrative through a process of interpretation and fantasy to their given details. The scripts were enacted by a selection of readers and composed into audio dramas. The final tracks sonically translate the memories back into sound, teasing out the embedded and imaginative thrust of the given memory.

What has always interested me in this project is how the memories continue to speak and I feel thoroughly inspired by all they suggest, from their highly personalized reflections as a form of cultural study to what radio memory reveals about listening and technologies of broadcast: might radio memories suggest a musicological perspective informed by questions of place as an emotional geography, or aid in charting out an ontology of transmission by weaving together experiences of listening with the very systems that put music into the ear? With this in mind I'd especially like to thank all the contributors for sharing their memories, and for considering the project worthwhile. I have elected to present the memories here as they first came to me, with all their given misspellings, typos, and grammatical flourishes – I feel that such things only add to the very personal nature of memory, and the distinctiveness writing adds to their retelling.

BOXING MEMORIES

Bastien Gallet

Writing as if I was listening to radio. Not any though. A very special broadcast. My (own?) voice speaking another language to my (own?) ears. Writing foreign words is listening to a radio transmission. Therefore. I write what I hear. Which comes from... who knows where? I used to know. For instance a studio in a radio station. For instance in Radio France. I spent so many evenings in front of a microphone (more than once a week from 1999 to 2004) I gave him a name, shall I tell you? Later. I spoke the words my mother taught me to no one I could imagine but the (wo)men I interviewed and the (wo)men I saw behind the window (and with whom I talked when no one could hear me but them, and they talked to me through my earphones when I was speaking to no one I could imagine). No one I could imagine makes a lot of people, sometimes 50,000 people, sometimes more, sometimes less, and meeting them afterwards, one and one and one of them, was not meeting no one I could imagine, was not meeting anyone, only weird and far images of what was on air this evening. My own voice but not speaking anymore and divided in two parts. Never united. Never one. People who listened to my radio transmissions told me about what I meant to say (a part) or what my voice was like (another part), meaning or texture, one or the other, as if they were two very different things – which I suppose now they are, for they are the result of the division radio performed on my voice. From which I conclude this: radio transmission does not separate one's voice from his body for we (anyone who listens) hear one's body and flesh and soul in his voice; radio transmission divides every voice in two parts and these two parts are equally the voice itself and its body and flesh and soul. Therefore, what streams forward, transduced into electromagnetic waves, are divided parts of oneself, as meanings and textures, which we always hear together but never at the same time... Is that why radio behaves as an impersonal memory? Is she? Voices. What about music? Music has no meaning, music is plane and flat and deeply flat, a river flow of textures and surfaces. Radio cannot divide music and that's why it always brings so many things with it, so many sparkling events, colored threads from outside, other places, other lives, non-musical materials and souls. Every piece of music heard anywhere is an open door, not always, if it encounters someone's life, if this life is tuned in to this piece of musical time – how do we do that? Nothing to do. It happens. What happens? I can tell you right now. Listen. It's my voice you're hearing, written by myself as I heard it as if she was coming from a radio transistor:

I was fourteen years old, learning how to sail on a small Breton island, what was its name, it must have been "L'île aux Moines" (Monks Island). I remember the ruins of an old monastery perched on the top of a cliff, we could hear near a certain wall still upright the crashes of the swell resonating across the stones and think it was the remote echo of the singing monks. It was New Wave summer in France, radio was filled up with Depeche-Simple-Fad-Cure-Minds-New-Tears-For-Mode-Order-Gadget-Fears and I was desperately in love with a curly blonde girl called Marine. We were sailing all day long in junky boats not easy to handle and moving unrhythmically at night to New Wave incredibly synthetic sound. Every night came the same song at the same time and everybody would sing the same words "shout, shout, let it all out" I can still hear them now "these are the things I can do without" whirling in my head "come on, I'm talking to you, come on" as Marine is being kissed by some guy I had never seen before. This is not the memory I'd like to share. Nobody would. Happily there is another one. Not very long after this sad incident, I climbed the Ruins Cliff at night, had a small radio my grandmother gave me for Christmas with a long antenna (she possessed a television but preferred to listen to radio transmissions of football matches), wanted to capture weird waves, the ones that only reach the highest points. There was no wind, only the sound of the swell. I spent two hours scanning all different types of frequencies, eventually got to shortwaves. When I heard it, the moon had sunk in the ocean, it was very dark, I was tired and thought it was another kind of interference. But it happened to be a voice, and not any voice, a prayer. A woman was praying (it was too litanic to be anything else) in a language I didn't know. Her voice was fluttering (of course it might have been an effect of the transmission) but there was a kind of joy in it, an ecstatic joy. I listened and listened and the voice

This short story is not completely true but I can't say which part of it, which memory never was a fact. A fact? Can a radio transmission be a fact? Yes, as long as it's not an event. For that's what it was: an event. More precisely. An intersection. Between two states of being, two states of thing, two planes filled with virtual beings. Memories on the one hear, sound threads on the other. A tuned in memory knits a thread and pulls it and knits with it every other thread it brings, which are as well other memories then knitted together. Well, these many threads, let's put them in boxes, rolling wood boxes hither and thither, yellow and blue and green and red, many boxes on wheels of steel filled with threads of that kind, ready to be knitted by anyone's memory, or sewed on anyone's body. Unconscious knitting resulting in a sudden awareness. I remember. I amplify. Everything vanishes except the song that is singing from my past. Memory is an amplifier. Memory is Lee de Forest's grid. In the middle of the electronic flux / in the middle of the radio flux, between anode and cathode / between me and my past, Lee de Forest (1906, patented in 1907) / God intersected a grid that controlled the intensity of the electronic flux / that controls the intensity of the radio stream. For there are two signals: the electromagnetic waves & the sound waves and both need amplification, both deserve a grid. Lee de Forest / God called his new device the Audion / Memory and claimed later he invented radio / human being (for his Audion / Memory, in spite of many not-from-him improvements could detect, amplify and transmit radio waves / things from the past).

Therefore. Radio is glossolalia, translates sounds and voices from one place to another and doing so puts in someone's ears a language he doesn't speak but he speaks nonetheless. Translated words need another kind of translation. In Geneva – Maison communale de Plainpalais, rue de Carouge, mars 2005 – where these boxes rolled on every floor, one heard the music, didn't understand the words, even when he knew the tune. I didn't understand the words of that song. I thought I heard – shout, shout, lady along – although I couldn't see what it was supposed to mean. My memory isn't like one's of a mother tongue's rememberer, it's of another kind, a signal without sense but many collateral memories, a character in a complex scene, one point in a constellation. Therefore. We listened these boxed (kicked and wrapped) memories but heard a tiny part of them, sometimes we would read one's story and think about it and tell to ourselves after awhile that the stories these songs were recounting or inventing were nothing compared to the stories they ought to be part of. Then we would look for the tunes that were as well a part of our memories and compare our stories to the ones that were related. We would start to remember things we had never lived, and bit by bit a few amidst us would become a few someone's else, strangers to themselves, suddenly understanding and speaking two languages as if they were both native. Pentecostal boxes. Let's push & pull them, let's make them roll, let's collect and gather a few or a bundle of them and mix their songs. Let's build spatial bootlegs. Let's cluster together all these memories. And let's imagine the memory of them all, the memory of all memories. How would we name it? Humanity? Humanity as a mere collection of memories? The name of a receipt: take all the memories you can get, put them in a big pot, cook a million years. Humanity is an assemblage of boxes. Radio is a fiction operator and a life mixer. Radio is the big writer, the greatest storyteller. Radio is the active intellect of mankind.

I'm talking to you. That's what they say. To no one else. This tune imagines me, any me and tells words to my ears that are for no one else but me. Another trick of the trade, another merchandise prestidigitation, another deceived consciousness... Could be. But am I not actually using this song, tricking it in some way, talking to me through its words and keys? Which part of me is exhorting the other? The ancient me young me not any longer me urging the now me old me me for sure? The being asking the non-being if he remembers what he cannot but fight and try to obliterate, what pushes him into the great nothingness of the future? Maybe. This song now makes me cry. Because I can't unlink it from its thereabouts. Even now it's in the box. It still blows my present. These boxes are nothing but a stock of explosives waiting for a consciousness, a life with a burning match. A memory is a detonator.

These imprisoned musical threads were thus put in boxes for every one of them had been remembered, amplified by someone's memory, sealed off as a single part of his past. These already marked songs and tunes, already knitted together with fragments of souls, there, in these boxes and emerging from these boxes, are claiming their differences from the others still buried in the radio flux. They have been unearthed. Two times at the same time, from one's forgotten past, from the indefinite present of the radio course. But now they are isolated two times, from the past and from the present, no more past and no more present are they, but not yet in the future. In what time are they running their not very long course? What kind of time is not, never was, never will be? None I know of. For their time is virtual and nevertheless real. Singular essences. Reduced memories but not in a phenomenological way. They've not become noemas. Only persisting objects. Focal points of any possible rememberance. Crossings. Powerful boxes that can produce such a virtual time. Who made them? Neither God nor Lee de Forest. Brandon LaBelle, his name is. Artist. Who else would have done that. To sing these songs in no time.

I nearly forgot to tell you the name I used to give to my microphone. Ferdinand the monk.

—

Bastien Gallet is a curator and writer from Paris. He has worked with radio, holding a weekly radio broadcast on France Culture, as well as acting as director of Archipel music festival, Geneva, from 2003 to 2006. He currently works as a publisher for MF éditions. He has published two novels, essays on music and an archaeological fiction about Rome (Anastylose). He teaches aesthetics and philosophy in the south of France at Montpellier's art school.

THE NON-SHAPED BEING

Carmen Cebreros Urzaiz

No other time like this, ours, has been provided with such a load of devices and systems to register the events in which we get involved. It doesn't matter how domestic or how grandiloquent the occasion to record: recording accompanies and synchronizes with happenings. There is an actual generalized spirit of anticipation which seeks to prevent oblivion by consigning methodically everything we do, everything everybody does. Such testimonies take different forms according to the technology and technique employed, and immediacy comes to impose without detriment to the sophistication of these technologies. In fact, the more immediate we need them to act, the more sophisticated they are, although we don't realize it since we have incorporated an array of technological devices into our daily practices: data files, instant photographs with mobile-phones, contact details automatically saved, massively distributable reminders, video and sound recordings, etc. Whatever happens, a parallel proof is constructed for the future, attempting to live both as one single fact, living one as the shade of the other.

We use these forms and codes implemented by recording technologies in order to share points of departure and reference with others. But isn't it that as soon as an analogous image is captured and presented to us, we start understanding the episode as that image and erasing the one in our mind, adopting the external information as if it were our very inner one?

Is memory the accumulation of such material preserved for the future? For what we imagine the future would be? Or is memory the actual remnants living as images in the mind of individuals with no material possibility of existence since any materialization would constitute the emancipation of the very memory? Is it that as soon as we attempt to reconstruct the past we start reinventing it?

I couldn't affirm that this corpus of material produced and stored to guarantee the possibility of a complete reconstruction of our existence is addressed for our own use. All this is done with the image of a seemingly anonymous character willing to reconstruct us. One could endlessly go through and reflect on this extensive urge to register and capture one's own experiences, but then when would we live and keep filing our lives?

Borges' Funes (the memorious) was a man who was able to remember every single detail of any image, sound or texture he captured, and he was even able to remember the number of times he had imagined and thought over every memory, or the combinations of them. He spent complete nights asleep in the dark, reconstructing, for example, the irregularities of a crack or the clouds he had seen as a quick look earlier in the day, or any other day. He tried to design a system for his mind to store his memories to the exact number of seventy thousand, but he noticed that his life, what was left of it, wouldn't suffice to make an accurate selection, since each time each memory would reproduce and expand in nuances. Funes was old at the age of nineteen because of the amount of memories he had accumulated, furthermore, for his capacity of reviving – and he did revive extensively – his experiences without hierarchy. Remembering and archiving represents then a dispute between appealing to the non-shaped within our minds and the codified within our archives.

Music, which is the historical result of a particular type of coding, recording and filing, paradoxically and dynamically infiltrates as an incidental device for our memory and acts of remembering. Songs and melodies randomly occupy the environments in which we evolve. We decide to keep record of things because we are not able to decide (and make it happen) when and how something will be fixed, for bad or good, in the loose space of our being; when might something happen that would carve us, having an inadvertent sound at the background.

This plethora of sounds, rhythms, melodies, words, phrases, songs get stuck in our minds without mediation of taste, and beyond overly intellectual intentions.It is coincidence and its repetition that incorporates these tunes into sense of self as signs and hints through which we are able to recuperate a complete scenario we were part of.

Time operates through us in strange ways. Maybe what is strangest is our fallacious idea of synchrony and the continuous present: that we are while we live, and that we are what we live. However most of what we live is not grasped as experience, and even less as significant "in real time". We are more capable of recognizing we were living, and indeed alive, via remembrance.

When we listen for the second, third or hundredth time to a song that accompanied an event, its repetition makes us remember, and furthermore to become aware, that we attended to that particular event which is our life: It constitutes our own emotional landscape, it is ours, it belongs to us, and it is relevant and current.

—

Carmen Cebreros Urzaiz is an independent curator and writer. She graduated from the MA in Curating at Goldsmiths College, London. She recently coordinated the project "Piano Recital. A project by Fernando Ortega" presented by kurimanzutto in collaboration with Casa del Lago Juan José Arreola and Radio UNAM, Mexico. She has curated the exhibitions "The Taming Power of the Small" in collaboration with the art-collective Proyecto A-Corpus (Mexico, 2003), "Stages and Transfers" (Mexico, 2005), and "Sir John Soane's Museum Audioguide Project" (London, 2006, in collaboration with Claudia Rodríguez-Ponga). In 2005 she was granted the Miguel Covarrubias Prize (National Institute of Anthropology and History, Mexico) for her dissertation "Contemporary Art Museums: Present Time and Society." She is currently teaching at the National School of Fine Arts (UNAM-Mexico) and is the archivist and registrar at kurimanzutto gallery.

TRANSMISSION 1: HAUNTED HOUSE

Where the quiet of interiors resounds with unseen murmurs, sudden ruptures, and disquieting echo. Televisions, radios, neighbor's voices, the passing of cars outside, the humming and gurgling of appliances, arguments, dinner chatter, whispers in the dark, all orchestrate a compositional ground of home, that in turn creates surprise – songs that puncture through, pop the bubble, insert sudden creatures.

Total Eclipse of the Heart, Bonnie Tyler

Solsbury Hill, Peter Gabriel

Changes, David Bowie

Rise Above, Black Flag

True, Spandau Ballet

Børne Radio Show

Radio Ga Ga, Queen

Never Mind the Bollocks, Sex Pistols

Rumours, Fleetwood Mac

Follow the Leader, Eric B. & Rakim

Clocks, Coldplay

string quartets, Shostakovich

Total Eclipse of the Heart, Bonnie Tyler

My parents had just separated for the final time. My father picked me up for a
rare outing and as we drove down Hawthorne Blvd on our way to who knows
what, he attempted to explain their separation. It was an uncomfortably intimate and adult conversation for the pre-teen that I was, never having experienced or even conceived of the love or lack of love he was trying to elucidate
– but there was nowhere I could go. He had recently purchased this cassette and
played the song for me a few times as it apparently expressed his feelings better than he could himself. The effect was long-lasting indeed. My heart was
eclipsed, too.

Contributor: Angela Zusman
Age: 34
Location of memory: Car, driving in LA
Year of memory: 198?

—

Solsbury Hill, Peter Gabriel

When I was a schoolboy, my mother listened to BBC Radio 2 every morning –
safe, uninspiring, background sounds. I was a dull teenager: fatherless, rudder-
less, clueless. Flightless. One morning, as I was leaving the house, I heard something different, and stopped to listen – I couldn't leave it, and I missed the bus
to school – and in the time it takes to conceive a child, the song had reached into
me and flipped the "music" switch to the "on" position. I remember everything
– the smell of the kitchen, the wallpaper, the weather – about those few minutes,
27 years ago, and they shaped my life. I became a listener, an enthusiast, a bore,
a collector, a gig-goer; and, as a direct result, a guitarist, a songwriter, a profes-
sional performer, a sound engineer, a tour manager, a celebrant of the intimacy
and passion that can be conveyed in a single act of musical communication...

27 years, 60 countries and several thousand gigs later, I can trace everything
I have and everything I am to those few minutes, so long ago, when something
came out of the radio and caused my life.

Contributor: Rog Patterson
Age: 43
Location of memory: Birmingham, England: the hallway of our family home
Year of memory: 1977

—

Changes, David Bowie

i had a schizophrenic father, and i knew many moments of loneliness. When I
heard 'changes' by david bowie, i suddenly understood the sense of feeling old
before my time, and that i wasn't alone in the world. this song somehow gave a
name to my sadness and made me feel less alone. I felt i had found a soul mate

Contributor: Denise Dee
Age: 46
Location of memory: Pittsburgh, Pennyslvania, the cement basement of my par-
ent's house washing my hair
Year of memory: 1971

—

Rise Above, Black Flag

Like many suburban male teens in the 1980s I listened to a lot of heavy metal,
most of which has come to be known as "hair metal". The standard issue Motley
Crue/Ratt/Twisted Sister fare satisfied, for the most part, my craving for music
full of aggressive energy, music that gave voice to the internal struggles in my

typical teenage mind and body. But there remained something cartoonish about these bands, even beyond the make-up, something that undercut their "dangerous" image. A local rock radio station, where I got my daytime fix of hair metal, had a Sunday night show called Critic's Choice. As a concept, this show existed in most American cities in the 1980s. It was an echo of 1960s and 70s free-form radio. In the 1980s punk and hardcore were the unexplored territories in commercial radio. My most palpable memory of that radio show is hearing Black Flag for the first time. It scared me. I had never heard such violent music, such real aggression manifested through sound. In barely more than two minutes Black Flag showed how calculated the "dangerous" image of hair metal bands was. I kept listening every Sunday night and heard more Black Flag, more hardcore, more of many types of music that would have a profound influence on me. I always listened with headphones on; my parents brushed off Motley Crue as a joke ("they're wearing eye shadow and blush" my mom would say), but Flag was real. They would be shocked (I hoped). Like a lot of teens, I also felt protective of this secret music. I wanted to hold it tight. Eventually I learned that "Rise Above" was more than three years old by the time I heard it. I clearly had a lot to catch up on.

Contributor: Greg Headley
Age: 37
Location of memory: Lancaster, Texas
Year of memory: 1984

—

True, Spandau Ballet

I remember having Spandau Ballet's "True" on my mind, after having seen the rather cheap horror movie "The Fog" late night on my parents' television. Though or because i was aged fifteen, i was not able to cope with this scene where zombies were approaching a lonely lighthouse from the seaside. My brain

replayed it over and over again, accompanied by this particular line in the re-
frain "...I know this...much is...true !!". I was scared as hell and didn't manage to
fall asleep. This vision and sound caused the very worse night of my life. In the
meantime i have developed a rather distinguished taste for music and i cannot
more than laugh about this, but the song keeps driving me nuts.

Contributor: Gunter Adler
Age: 35
Location of memory: My parents three-room flat in Aachen, Germany
Year of memory: 1983

—

The radio program was called Børne Radio and was hosted by childish grown up radio hosts and well-articulated and funny children at my own age at the time.

The first and only time I got hooked on a radio program was when I was 11-13
years old. It was a spooky and imaginative serial that children and the hosts
elaborated on every day for an hour in the afternoon. The frame of the story
was historical and started of in the middle ages and evolved to present time over
the years and dealt with issues interesting for young children. The stories had
to do with the ups and downs children face with mates, parents and themselves
and was done in a most engaging way with loads of sound effects and a lot of
sarcasm, irony and humour. Sometimes the children phoning in would get a
hard time by the hosts if the story wasn't innovative or interesting enough and
they would cut them off the line. I spend at one hour a day listening to the program for approximately two years. It is the only radio serial I have ever been
involved in.

Contributor: Malene Rørdam
Age: 33
Location of memory: I listened to the program in my room, where we lived in

a flat in Lyngby (a suburb to Copenhagen), mostly when I was doing my homework or drawing
Year of memory: 1982-84

—

Radio Ga Ga, Queen

I was a child visiting my cousin, he had this strange music that I didn't like, but
somehow understood was something new and meant something to my world.
My cousin said it was new and a hit and very good, perhaps therefore I listened
more, – in any case he played it over and over and I started to like it. I always remember this afternoon when I now hear the song, the afternoon when I started
slipping out of childhood.

Contributor: Michael Madsen
Age: 33
Location of memory: In my cousin's room
Year of memory: Perhaps 1980?

—

Never Mind the Bollocks, Sex Pistols / *Rumours*, Fleetwood Mac

saw Telly Savalas on some TV awards show present a short report on The Sex
Pistols. the next day I bought Never Mind the Bollocks and, for good measure,
the big winner that year, Fleetwood Mac's Rumours. this happened shortly after
my dad, my brother and I had moved to Canada after two years in Argentina. in
Argentina I hadn't really been aware of pop/rock music. Before that, in Geneva,
as a child, I was into French pop and the Beatles, but it was really through my
mother's record collection. so that Telly Savalas moment felt like it was bring-
ing me into the present and that soon developed into a weirdly schizoid musical

taste palette consisting of punk/new wave, progressive rock (Québec has always had, still to this day, a huge fan base for Genesis, Gabriel, etc.), and let's not forget disco. Well, disco kind of faded, progressive rock became embarrassing, so all that was left to do was to go deeper into punk/new wave which eventually emerged into all things experimental. As for Fleetwood Mac, took me a while to realize that Tusk was the much better album.

Contributor: Christof Migone
Age: 40
Location of memory: Montréal, home watching TV
Year of memory: 1978, um... it could actually have been 1977, not quite sure. we moved to Montreal in Spring 1977

—

Follow the Leader, Eric B and Rakim

I was around 16 years old and was getting ready to go to school, taking a shower, etc, with the radio on a soul/r and b type station which sometimes played hip hop music, which was in its infancy at the time.
I heard Eric B and Rakim's "Follow the Leader" which had really fast , interesting rhymes in it. I remember that I was living at my fathers house at the time, rather than my mother's house. So it was a different room and space for me to wake up into.

Contributor: Olivia Block
Age: 34
Location of memory: Dallas, Tx, USA In the bathroom of the house I lived in with my father, stepmother, and step-siblings at the time
Year of memory: 1996

—

20

Clocks, Coldplay

I was in a bus that was going from Heathrow Airport to Stansted Airport, in London, England. The bus driver had a portable radio and as I sat in my seat, I heard "Clocks" by Coldplay. I still remember smiling when I heard "home, home, where I wanted to go".

Contributor: Nadia Bartolini
Age: 30
Location of memory: Bus from Heathrow to Stansted
Year of memory: 2004

—

What i hear in the background was string quartets no 7 (Op.108) and 8 (Op.110) and piano quintet (Op.57) executed by Borodin String Quartet with Sviatoslav Richter.

though my modest collection contains many beautiful recordings, this one still 'hurts'. my radio memory is not a radio memory but the experience had this special impact you are asking for and it came to me in an almost-radio-way. it was during an intimate talk with a woman i was in love with. during our phone call, the music in the background of her voice seemed to follow our conversation and made me cry. it was a recognition of beauty, and my mind easily wrapped around / became occupied by the composition and the interpretation of it. next day i bought all compositions for string-quartet by Shostakovich as performed by the Borodin String Quartet and stayed in bed for a week. Listening.

Contributor: Melle Hammer
Age: 50
Location of memory: amsterdam, sitting in the doorstep to the balcony of my apartment, phone in the right hand / a glass of wine in the left
Year of memory: September 1996

TRANSMISSION 2: SILENT NIGHT

The paleness of starry skies... shapeless forms, shadows of wind and dew on still fields, or streets caught in their own sleep. Where time stops and space forgets itself, revealing the eyes of forgotten souls, ghosts of childhood days, drifting spirits finding their way on the vapor of lyrics, melodies and dry rhythms.

Ol' '55, Tom Waits

Incoming, John Duncan

Wonderful Tonight, Eric Clapton

All Night, Sam Phillips

Missa Aeterna, Giovanni Pierluigi da Palestrina

River Man, Nick Drake

In Your Eyes, Peter Gabriel

Ol' '55, Tom Waits

The song came with the dawn, on dewy fields, as the embers of the fire and the party faded. It was the end of a college year, and the summer and the excitement of a possible love were beckoning. Someone must have loaded a tape and pressed play. In my mind the sound comes with the sun, reaching across the fields.

Contributor: David Pinder
Age: 34
Location of memory: Fields by a river, cows chewing nearby, on the outskirts of town
Year of memory: 1993

—

Incoming, John Duncan

Standing out alone on the roof terrace of my house, watching the slow progress of a lunar eclipse. At the end of a tortured 10-year nightmare of marriage, depressed, slightly stoned and drunk, looking for the energy to celebrate my birthday. Experimental music from the pirate radio station Radio 100 filling the house: 'Discipline', produced and hosted by my friend Peter Fleur. Exactly at midnight the moon reached full eclipse -- and at that same moment I heard my own music ('INCOMING') over the radio. It defies all rational explanation how Peter, broadcasting blind from the windowless concrete balcony I knew so well from making my own broadcasts there, was able to synchronize the music perfectly with the eclipse, how he could possibly have known how much it would mean to me to hear my own work by sheer chance at just that moment, to know that he thought enough of it to broadcast it of his own free will. Probably he had no idea of any of this, which makes this memory even more special. I remember it as a kind of epiphany, a tiny sign, utterly trivial to anyone else, of motions well beyond anything we can imagine, a lasting encouragement to continue meditation as much as to keep making music.

Contributor: John Duncan
Location of memory: Amsterdam, Red Light District
Year of memory: June 17, 1995

—

Wonderful Tonight, Eric Clapton

the year was... 1994... i was 18 at the time, the song i heard was eric clapton, "wonderful tonight". i was in this clinic, a waiting area before going into operating room. there were these other girls, some my age, some younger... nobody knew each other... a nurse comes in and takes one person out to the operating room every 5 minutes or so. and she'd put these i.v. tubes in our arm to prepare to go under. this one girl had problems finding vein and she began to cry when the nurse was probing her arm and hands for a vein. then, after getting it plugged in, she was moved next to me, then my arm next... the girl kept weeping, it made me weep,and i noticed some of the other girls were weeping, no mothers allowed at that point.. everyone is alone... ... that song comes on with those lyrics.. made me think of my boyfriend who loved eric clapton... that song... me... and i began to cry so hard seeing all of us in there, waiting to get abortions and i wished so much that he was allowed to be in the room with me at that time. instead, the girl next to me who was crying, we held hands and said good luck before she left the room. and still when i hear that song, i remember that room...

Contributor: Anonymous
Age: 28
Location of memory: Louisville, Kentucky...
Year of memory: 1994

—

All Night, Sam Phillips

I had been dating a girl for a short while when I decided to break things off for good. We were standing in her apartment having a serious discussion about why I didn't think things were going to work out. She had a small radio in the kitchen playing music, though the large room and high ceilings made the cheap radio sound tinny and distant. During a thoughtful pause in our conversation, I heard a song I'd never heard before come on the radio. I had not been particularly enjoying our conversation, as few people in that position do, so my mind may have been particularly interested in drifting. The song immediately overwhelmed me, and I could barely concentrate on our conversation, even to the point where I interrupted my friend and asked her if she knew the name of the song (she didn't). The song's sultry voice and inviting rhythms were soaked in the kind of passion and yearning that was missing from the relationship I was in the middle of ending. I continued to listen closely to the radio for several songs more, until the DJ came on and finally identified it as Sam Phillips' "All Night." The song remained in my head that night during the long drive back to New York City.

Contributor: Peter Krause
Age: 34
Location of memory: Catskill Mountains, New York
Year of memory: 2004

—

Missa Aeterna, Giovanni Pierluigi da Palestrina

It was dark and candlelit in the basement of my mother's house in New Jersey, the house built by my father and my grandfather forty years ago, the house where I grew up. It was Christmas night 2001 and my wife was in labor with our son Lio. To help ease the labor pains and welcome our child into the world as peacefully as possible we had various early modern, Renaissance and early-modern-like music CDs playing (William Byrd, Hildegard von Bingen, and John Tavener). When Lio was born amidst water and smoke Palestrina's "Missa Aeterna" was playing. Everytime I hear it I remember that happiest moment of my life. Although now that I've written it down and read it back it sounds a bit clichéd, a bit of a middle-class fantasy. But the fact remains that I'm much more fond of this memory than of the memory of my losing my virginity to Rush's "Tom Sawyer."

Contributor: Martin Spinelli
Age: 37
Location of memory: In front of the old fireplace in the basement of my mother's house, Martinsville, NJ
Year of memory: 2001

—

River Man, Nick Drake

I was an awkward teenager, not especially social, not from a lack of desire but from a shyness that was painful to escape. I would often sit in my bedroom for some hours listening to the Radio London, since shut down, a station that offered up rare delights of esoteric beauty. It was on a Sunday afternoon, the soft tones of the announcer introducing at length in a drowsy breath heavy manner the songs to pass the day. My mother came to my room to explain that my beloved grandfather had just passed away after battling with an especially ferocious type of cancer. I adored my grandfather, he'd brought me up as a child and my happiest memories stem from this period. She closed the door and I sat there, hollow, quiet, lost in my thoughts, and Nick Drake's 'River Man' began to play on the radio set in the corner. This richly elegant voice seduced me, the delicate string arrangement enveloping the melancholic ambience and in many ways I slipped inside the music, tumbling over the edge of reality, following a spiral staircase of mourning in silence. Nick Drake's songs have continued to offer a mirror to these moments in my life. Whenever I hear this song it returns me to this quiet room, this heartbreaking moment in time.

Contributor: Robin Rimbaud / Scanner
Age: 36
Location of memory: My bedroom, Southfields, London
Year of memory: 1978

—

In Your Eyes, Peter Gabriel

Finding out that my boyfriend (and the man I was in love with) was actually seeing another woman. On the radio, in the background, "In Your Eyes" by Peter Gabriel was playing.

Contributor: Nadia Bartolini
Age: 30
Year of memory: 1998

TRANSMISSION 3: PARTY MUSIC

And all hell breaks loose! With teeming bodies, rapturous momentum, the fractured days finding their form in nights of collective euphoria–strangers ecstatic with rubbed elbows and sweat-tickled kisses: where DJ and crowd coalesce, coagulate the dripping tedium of passing moments into pockets of gathered fevers. Lost and found in the same moment...

Livin' on a Prayer, Bon Jovi

Losing My Religion, REM

Jump Around, House of Pain

Little Wing, Jimi Hendrix

Fascination Street, The Cure

Prologue, Gentle Giant

Fast Car, Tracy Chapman

The Whistle Song, Frankie Knuckles

Candlestick, The Cannanes

Heartbreak Hotel, Elvis Presley

Centerfold, The J. Geils Band

Atomic, Blondie

Don't Look Back, Boston

Goodies, Ciara featuring Petey Pablo

Vaya Vallende, Carlos Gardel (Amargura) & Celia Cruz

Would you like to swing on a star, Bing Crosby

America, Neil Diamond

Losing My Religion, REM / *Jump Around*, House of Pain / *Fascination Street*, The Cure / *Little Wing*, Jimi Hendrix / *Living on a Prayer*, Bon Jovi

I was at a beach party in Rimini, Italy. Everyone was dancing except me (I was at the bar). The music was all Italian pop-rock, which I knew nothing about. So I drank. Until I heard a familiar tune. With a Martini in one hand and 3 olives in the other, I pushed my way to the dancefloor and danced like a madwoman screaming the words to "Losing My Religion" by REM. Finally, some music I could relate to. 1992. – Totally hammered in a bar after I started my first week of University. "Jump Around" by House of Pain came on the speakers and everybody started going bananas. I remember being hit from all sides, and making out with a freshman as we were being showered with cheap beer. A moment I would rather forget. But I still like to jump around when I hear that song. Ottawa, Ontario. 1992. – I was in an alternative-goth bar in Montreal. Met a fascinating man, whom I followed to his apartment (the walls were all painted red... yikes). When we first spoke at the bar, "Fascination Street" by The Cure was playing. How appropriate, I thought. 1997. – The first bar I went to in New York City. I stayed at the bar the whole night drinking rum and coke. At the end of the night, the barman came up to me and said: "I'll ask the DJ to put a song for you – when I saw you, I automatically thought of one song". The DJ played "Little Wing" by Jimi Hendrix. 2001. It's still my ultimate comfort song. – Another bar... this time in St. Petersburg, Russia. The first song I heard when entering the bar was "Living on a Prayer" by Bon Jovi. Wow. I thought I had just entered a time machine. All I could think of was "Beam me up, Scotty". 2003.

Contributor: Nadia Bartolini
Age: 30
Location of memory: Mentioned above
Year of memory: Mentioned above

—

Prologue, Gentle Giant

In 1985, I had the experience of two new worlds harmoniously fusing, with the result being that 'my world' began again, and was never the same since...I was at a friend's beach house in southern New Jersey, and was introduced to, for want of a better term, a 'psychedelic happening.' Three friends had gathered to ingest a rather powerful mind-altering substance. To accompany this procedure, my gracious host created a soundtrack that became perhaps the most exciting musical experience I ever had. The music was by the adventurous British band named Gentle Giant, the album was "Three Friends," and the decisive track was the opening "Prologue." Things were being done-- worlds were being explored in this sonic gesture that I could never have imagined. I took it as an invitation to grow, and realize the wonderousness within me. My mind was absorbing visions and possibilities that I am still exploring today. Meanwhile, outside of my overworked brain, my host was dancing maniacally to the music with a teddy bear, helping me to laugh at this aesthetic overload, as profound as I thought it was. Bless you, Gentle Giant, and the divinely imparted material resources that can thrust one into new states of being...

Contributor: Andrew Blakeslee
Age: 34
Location of memory: Beach House, New Jersey
Year of memory: 1985

—

Fast Car, Tracy Chapman

This song was playing as a carload of us returned to the Greenpeace headquarters after a night of canvassing membership. Driving home was definitely a highlight of a summer job going from door to door selling to unsympathetic – and harassed – householders. This night the sun was only setting, midsummer, and we were skirting a city bay; one of the young, handsome, passionate and naïve canvassers had been sleeping and woke to ask whether Tracy had been singing to him in his dream or in real life. The song is as romantic as the people I was working with; later that summer we would all be laid off as Greenpeace ended its door-to-door campaigns. I wonder how wise it ever was sending ill-informed fanatics out alone in the evenings to interrupt the dinners of the people of Massachusetts, but this thought and my perception of that time has been much clarified over time; in 1997 I was a square peg in a round hole and liked the song.

Contributor: Aoife O'Brien
Age: 26
Location of memory: Driving through the outskirts of Boston, MA
Year of memory: 1997

—

The Whistle Song, Frankie Knuckles

I grew up in the middle of nowhere, practically, so trying to find new & exciting music required quite a bit of effort. one source was the college radio station that broadcast from the university of Massachusetts, a 40 minute drive south of where I lived. there were one or two shows I would listen to late at night, after I was supposed to be asleep. these programs played pretty standard, eclectic college radio fare. I would lie in my bed with a tape player/radio permanently positioned between my pillow and the headboard and occasionally record parts of these shows. I had never heard anything like "the whistle song" before. I've heard it a few times since,(most notably a year or two after I first heard it, the song was used in a nestea commercial, the first time a song that I thought no one else in the world knew was used to sell something) and it strikes me as a great house track. moronically happy & bouncy, almost in a teletubbies kind of way, with a cliched "jazzy" build-and-release keyboard solo, but a great track.

Contributor: Ben Block
Age: 28
Location of memory: my third floor bedroom, late at night, town of gill, in western massachusetts
Year of memory: 1991

—

Candlestick, The Cannanes

It's a radio within the radio within the radio story. Within weeks of starting college, and having just met, my friend Miriam and I realized our only ambition was to have a show on the student radio station. We had no experience but were convinced our mix of experimental indiepop and eclectic world music would change the order of the universe. And we were given the 4 am slot on Wednesdays. This is the song I used to play to wake us up, at dawn. Miriam had never heard of the band, and I had only recently heard the song–on an anonymous radio show on the same station, days before starting our broadcasts. It sounds so pastoral and unaffected, but if you listen to the lyrics it's actually full of bitterness and sarcasm. It made both of us smile, groggily – all the out of tune singing and off beat drumming is the sound of smiling at 5:30 am. I've heard it on the radio since and feel that within there is a code that can only be understood by Miriam and me, wherever she may be I hope it still makes her think of every single one of those sunrises.

Contributor: Lupe
Age: 29
Location of memory: Swarthmore, Philadelphia, PA
Year of memory: 1993

—

Its 1956 – Reading in the south of England, about 35 miles from London to the west along the, what is now the M4 but then was just the plain A4. It is probably spring although you wouldn't really know that because its cold and probably wet. At least the windows of the bus i am sat on are steamed up with condensation. I am on the top deck of a double decker bus which is full of noisey teenage school kids, my classmates. I am 14 years old and i am suffering from culture shock and tropical nostalgia and wanderlust withdrawal symptoms. I am recently returned from Australia a place where my parents had decided to emmigrated a few years earlier taking me and my younger brother with them. Failing to settle they then decided to return to England. So here I am back where I started, where i had never started in fact. I only started on the six week voyage out, despite being so ill with seasickness that they almost put me ashore before we had arrived. Sat on the top of this window steamed noisy bus i am able to reflect on a trip down the Thames and into the English Channel, across the Bay of Biscay towards Spain and our first stop the Rock of Gibraltar and the first exotic soil beneath my feet. Across the Mediterranean to Marseille and my first realisation that i actually couldn't speak French despite having learned it for two years. How come they didn't understand? The Bay of Naples I do not remember ever seeing but apparently we stopped there; Several visits since do not jog the slightest memory whatsoever. But the smoking cone of the volcanic island of Stromboli does as does L'Aventura the Antonioni film. Port Said was a forbidden landing place. Nasser didn't like us British. We had to be content with purchasing camel skin wallets from bum boat salesmen who threw ropes up to us on the decks and passed goods up in little baskets in exchange for money passed down in the same basket. An element of trust involved obviously. A trip down the Suez Canal and the image of a huge passenger liner traversing the desert must have looked very bizarre from a distance. At the fay end of the canal was Aden then a British colony and army outpost now part of Yemen. My first desert place. Along trip across the Indian Ocean revealed flying fish, sharks and dolphins as well as water so blue I couldn't believe it until we reached Ceylon as it was then with its streets so full of rubbish I couldn't believe that either. But I could believe the Buddha that I saw for the first time and the music and temple dancers and the colour and noise and the long long white sand beaches and palm trees.

I loved Australia and the time we spent there. It never went away from me. But suddenly there I am on this bus with steamed up windows as if none of the above has happened. Where are we going? We are going for swimming lessons. These take place in an open-air Victorian public swimming pool situated quite near dear old father Thames. The water is freezing. I hate it. On the return trip a miracle happens. A formidable part of the class I am in are from a poor area of Reading renown for violence – The Orts Road. Most of them are budding Teddy Boys with a huge greased qiff of hair achieved with a deft flick of the wrist and a good plastic comb. Suddenly this part of the bus – the front six rows burst into Heartbreak Hotel substituting "Down at the end of Orts Road..." for "Down at the end of lonely street...". I am woken from my nostalgic gloom and thrust into rock and roll for the first time never to return. Only many years later did i become a reluctant swimmer and virtual surfer (in the warm waters of Tahiti at the age of fifty in fact) and cheap hotels are still an obsession

Contributor: Mike Cooper
Age: 62
Location of memory: Reading, England
Year of memory: 1956

—

Centerfold, The J. Geils Band

i was very young, driving in the car with my mom and next door neighbor who was the same age at the time and that song came on the radio. we asked what centerfold meant and my mom told us it was a bad word and not to sing the song... of course we continued to sing it after that!

Contributor: Rebecca Reeves
Age: 33
Location of memory: trevose, pa
Year of memory: i guess when it came out in the early 80's

—

Atomic, Blondie

Recently I heard Blondie's "Atomic" in a gym locker room I felt transported
instantly back to the early 80's when I was a teenager, to a late night in Boston
and the infamous 1270 Club!
It was a long, rainy night, and I am sure I was somehow impaired and acting out
like many of the other revelers. I remembered slam dancing before and after the
magic of this gem of a pop song made me freeze with amazement. That line "Oh,
your hair is beautiful, tonite...." just still sends chills. I remember losing the sole
of my boot that night, and accidentally ripping a whole swatch of fabric from a
big bruiser's cotton polo shirt. One evening, I was all powerful. Somehow, while
showering after a long and sweaty workout seemed like an awkward moment
to have memory serve me so vividly – but delving back into all this made me
realize that I had a mirror I would have had more than a twinge of an evil grin
on my face.

Contributor: TJ Norris
Age: 39
Location of memory: Boston, MA
Year of memory: 1981 approx

—

Don't Look Back, Boston

When I was a student of junior high, Boston's "Don't Look Back" was big hit in
Japan. We could listen everyday on the radio.
I liked it, but at that time me and my friend loved British rock music. We felt
American pop music is light and easy compare with British one. Sometime they
made us sick... for example, when we saw the make up and costume of "Kiss".
(Sorry....I don't think so,now. J)
We never trust with American pops. So I didn't buy their records.
Long time gone.... I have their CD.
When I listen it, I could feel atmosphere of my junior high period.
It was one of my happy time.
Now I am working at junior high school as a teacher.....
26 years past, Boston becomes one of my favourites.
"Don't Look Back"
They sing for me...???

Contributor: Toshiya Tsunoda
Age: 40
Location of memory: Yokosuka City
Year of memory: 1978

—

Goodies, Ciara featuring Petey Pablo

Last night was a great night hanging out with my friend. I haven't seen him in
a long time and I don't see him very often. He was really excited about this song
playing on the radio. I can't wait to hear it again ... I don't know who the artist
is ... but I'll recognize it ...

Contributor: Ulrik Heltoft

Age: 31
Location of memory: Last night: Los Angeles, driving (passenger seat)
Today: Los Angeles, waking up (hotel room)
Year of memory: 2004

—

Vaya Vallende, Carlos Gardel (Amargura) & Celia Cruz

On my birthday this year, I had nothing in mind; no party, no get together, no plans, no expectancy but a whole day reserved in which to play. I was surprised in the afternoon by a friend calling and inviting me to walk along the seafront to enjoy this day in high summer. We strolled, we pit-stopped for cocktails, we wound inland through the narrow cobbled streets of Regency Brighton, 'the Lanes', full of peculiar salt-sunken shops selling guns, jewellery, rarefied underwear – and poky cafes piping out voluminous fumes of hot chocolate. Then with a gentle tug at my sleeve, we slip left, west, the sun in our eyes and duck under the incongruously bright yellow and red awning of 'Casa Don Carlos', a Spanish tapas house. It's no bigger than a bedroom yet it's always crammed heaving with happiness, people flowing over with wine and paella and salpicon, mariscos, patatas fritas, platters and clatter and chattering all as loud as the space will house. We try to enter this capsule of noise – it's early but there are already people bunched up at the entrance waiting for a table. My friend is determined to feed me as a birthday treat, a swift surprise, so we loiter, shove a little, capture the eyes of the waiters hurrying scurrying. Then at once we are summoned in an imaginary andalé of the arm, and we allow ourselves to be corralled by the matador's gesture, the twittering maracas of the flamenco hovering like cicadas above a hubbub of food and summer skin humming in close proximity. A split-second lull. Then... ah, yes, that, what's that? I know this song – it's in my inner jukebox! This is from Colombia, I heard this at full volume on speeding buses racing me to work, and this is the song from that funny night in Marielita's in La Candelaría...

It's a sweet serendipity, the only time I've ever heard a Colombian song playing somewhere unexpected in England, catching me unawares. From sea to lanes to Spanish cabin to downtown Bogotá in a series of small steps, a flip-flop of hot toes and balls rising and falling, of cheap plastic on hot asphalt, melting vinyl. The arm of the stylus casts back to a dusty untouched tune, and the record whirls round in sympathy with memory, melody as memory in my inner jukebox.

So I lose the first few minutes of my birthday feast to a song; I sit in a tight corner listening to its story, remembering its fine details, hearing it as much as seeing it, led by sound.

Contributor: Colette Meacher
Age: 32
Location of memory: Brighton – Bogotá
Year of memory: 2004

—

Would you like to swing on a star, Bing Crosby

Of singing "Would you like to swing on a star" (Bing Crosby) when I was 4 or 5 (?) to a Future Teachers of America chapter in Stillman Valley, Illinois. My cousin Joan was a member, and I often visited my aunt's farm there from Rockford, Illinois. Our family listened to the radio all the time. My mom was a big Cub fan (I've written a poem about that memory), we sang a lot, and some of my girl cousins (the 2 who were my age) and I would put on shows. We laugh about one in particular: when we played Betty Hutton, Dorothy Lamoure, and Veronica Lake in the song, "The Sweater, the Sarong, and the Peek-a-boo Bang" for our families. We were a big hit!

Contributor: Nancy Harvey
Location of memory: Stillman Valley, IL

Age: 68
Year of memory: 1941 or 1942 (?)

—

America, Neil Diamond

I remember being very young, and my parents had a station wagon, a metallic
rusty red-orange color with chrome piping and trim, those crazy hubcaps with
all the chrome rods emanating from the center outwards in a latticework array,
and that carmely colored interior, dusty smell inside – this was back when my
dad still smoked (Marlboro's) – spare tire inside in the back, by the bench seats,
and we were driving south on West Jefferson, my mom driving – exactly-brown
hair cut short, green eyes with the tiniest tinges of hazel, thin elegant fingers,
eyebrows just right, and pretty cheekbones...and perhaps wearing that bright
fire-engine-red shirt with the white detailing around the collar and buttons...
or maybe just a faint blue knit cotton shirt – near the Detroit River in Trenton, Michigan, in view of the coal burning power plant with two white and
red ringed smokestacks, and the huge fuzzy black coal pile adjacent, it was
overcast and I was curled up and twisting around and about on the floor on
the passenger side of the car, nearly under the dashboard, singing along to Neil
Diamond's "America" on the radio, when we came to a stop at the light, waited,
and continued straight, past a small gas station and two askew smallish wooden
billboards, train tracks to our right, running parallel...late morning...

Contributor: Brian Taylor
Age: 27
Location of memory: Trenton, Michigan, under the dashboard of a circa 1980
Chevrolet station wagon
Year of memory: well, perhaps 1980

TRANSMISSION 4: HEARING THINGS

Buried in transistors and circuit-boards, the satisfaction of discovering, upon closer analysis the voices of other beings, the shadows that dance across walls or transcend the material plane: shape-shifting and charmed by sonorous events. Alchemy meets engineering meets giddy ears, alive with uncertain groping, charmed by knowing that something awaits, behind the surface, on the pillow or on the couch, driving through darkness or light of the hopes of a possible electronic home.

Reach Out (I'll Be There), The Four Tops

Difficult to say

I am a Baby (in my Universe), Daniel Johnston

none specific

Le Boeuf Sur Le Toit, Darius Milhaud

n/a (Radio Bulgaria)

I Stole Your Love, Kiss

entirely unknown

Lay Down Sally, Eric Clapton

Take Five (Live), Dave Brubeck Quartet

Elvis Presley

Reach Out (I'll be there), The Four Tops

My memory was sparked by the strange sound of an organ which introduced the song. I looked up behind me, to my left to the shelf where the radio sat. I was struck by the strangeness of the sound which was accentuated by the tinny transistor radio. It crystallised the memory. I was lying on my back on the living-room floor. My mother was in the kitchen directly behind me. The coal fire was burning a few feet to my left. I remember the cold grey light which came through the window. I think I even remember the colour of the radio (ivory coloured leather) though I can't be sure if I'm imagining it. The other details are correct.

Contributor: Steve Hunt
Age: 40
Location of memory: 3, Bellwood St, Harpurhey, Manchester 9 (since demolished)
Year of memory: 1966

—

Difficult to say

We lived in a military suburb housing area called Royal Oaks, 20 miles north of Madrid. My father was in the U.S. Air Force stationed in Torrejón Air Force Base, Madrid, Spain.
On my 8th birthday, October 27, 1959, my parents threw a surprise party for me. It was not so much the party but an unusual gift that has left a life-long impression. After blowing out the candles, cutting and eating a slice of cake with ice cream, my father and mother anxiously handed me a bright orange, black and white striped package with a small orange bow. As I peeled away the wrapper I could hardly contain myself as I got closer to the mysterious contents. From an unmarked, partially worn cardboard box I pulled out a black rectangular box

that had a silver grill at the top and several knobs. There was also on one side a dial face and small switches marked with text I could not read. I was astonished when I discovered what it actually was, not a baseball glove or chemistry set as I had hoped for, but a well-used Russian made radio. As I gained my youthful composure, I thought to myself, "what in the world was I going to do with this thing." It was sort of interesting but I had no idea how this cold ugly device was going to change my life. I looked up at my parents and smiled with an unconvincing, "thanks." It was not long after this disappointing event that I found myself obsessed by the strange radio device. We had a black and white television, which I seldom watched as the programming was limited to military news, football, and minimal Spanish programming. However, the radio became one of my best friends. That next year I was listening to it every opportunity I had to the various stations that could be found in that part of Europe – the nighttime came to life and was the best time to listen. Besides the BBC and Armed Forces Radio Service there was little or no English stations; however the non-English stations and the noises, static "in between" stations captured my devotion. I would immerse myself in the strange voices, music, and radio theatre and news reports for many hours at a time. Late at night after everyone else had gone to bed I stayed up with the radio hidden under my bed covers listening with an earphone so as not to disturb the rest of the household. One cold February night, after everyone had gone to bed I was listening for the heaving breathing and periodic snore bursts that would come from where my father and mother slept. I ritualistically turned the radio on and started exploring the radio airwaves beginning far left on the dial as I could turn, then slowly moved the dial to the right listening to the various stations and the rich noise between channels. I had just placed the radio on my bed near my pillow, when I discovered an unusual sound – a buzzing, singing noise with interspersed pops and clicks. A voice was there just underneath the noise; speaking English with a European accent, but the words were barely understandable behind these abstract sounds. Every so often a word would slip through that I could barely understand – especially since I was just 8 years old. I only have vague memories of the exact words and sounds. The earphone was not good quality and designed to fit in only one ear. I decided on this night to do without the ear device and just turn the sound down so as not to arouse suspicion in the rest of the sleeping house. I placed my head as close to the radio speaker as I could to listen for all the subtleties coming from this strange box. I found that if I put a wire or aluminium foil on the small antennae that I could increase the reception quality of the radio. One night I had extended a wire out of my bedroom window dropping it as far away as I could onto the ground to see if I could get this particular station to come in more clearly. The sound was still not clear even with all the additional metal attached to the antennae. I started turning the sound up little by little to see if I could garner what it was this person was trying to say. The voice was not a normal dj or newsperson. It sounded more like an every day person, possibly a preacher, or someone being interviewed talking somewhat anxiously with conviction and intent. After a while I became a little bored but wrote down the coordinates of the station to return later to this point to see how it was progressing and continued to explore further down the channels. Some time had passed as I explored other areas of the dial then came back to around the place where I had initially heard the strange voice and noise. The voice was no longer there but I continued to listen for some time, then dozed off maybe for an hour or so, possibly more. I woke up out of a strange radio dream space feeling kidney pressure. Forgetting about the radio, only thinking about relieving myself from the water pressure, I stumbled up out of bed making my way towards the bathroom. As I opened my bedroom door a strange voice from behind me whispered, "....these decisions to follow your desires will haunt you forever...." These words came out from the far corner of my room. I was frozen in my tracks, a cold chill fell over my whole body, the hair on the back of my neck raised up-- all the air in my lungs expanded suddenly causing me to "jump out of my skin." Who was in my room? And at this hour? I slowly turned towards where I heard the strange voice and tried to see in the darkened room but could not see anything except for the silhouette of some pictures on the wall, my dresser drawers, the bed and the subtle glow of the radio-- the radio of course.....I had left it on. My heart rate slowly returned to a more normal pace. Slowly, I moved towards the radio and the voice came back again a little less clear, but not as loud, then faded into another sound. I stepped back away again towards the left, then the right, and the voice came back more clear and louder, fading in and out, along with my movements. I rocked back and forth and the voice and radio noise responded to my gestures. I moved my arms back and forth, up and down, making the voice appear to interact with the noise, causing the voices to shift between another channel of what sounded like music from Hungary or possibly Bulgaria. I was thrilled about this new discovery that my body was extending and shaping the antennae reception. I was dancing with the radio at 4 a.m., totally enthralled with this new power. This went on for possibly 10 to 15 minutes, when, all of a sudden, again the hair on the back of my neck stood up with a new strange trepidation. I thought to myself, "....what now!" I slowly turned towards my partially opened door where just beyond the door frame stood two figures in light underwear watching me with a strange look on their familiar but twisted faces. It was my mother and father staring at me in bewilderment as they watched my early morning radio performance. My mother spoke first under her breath, "...are you ok?" Then my father added, in a slightly more stern voice, "What are you doing?" I was speechless how was I going to explain my behavior at this hour? In a small voice I answered, "Look what I can do!" I moved my arms back and forth in a flying gesture and the radio responded. With some relief, they both said, "That's nice dear, please turn off the radio and go back to bed, it's late."

Contributor: Steve Bradley
Age: 53
Location of memory: Madrid, Spain; bedroom
Year of memory: 1960

—

I am a Baby (in my Universe), Daniel Johnston

Listening to a radio program called "Corpse of Milk" on the college station WIDR, Kalamazoo, MI. I used to listen very quietly under the blankets from

12 – 1 am on Sunday nights during early high school years. This show often combined absurd material with the macabre or obscure, found items, cheesy lounge records and the like. One night I heard a song and in my half-asleep state vividly imagined a wrinkled, hunched over old woman, singing to herself under the stairs of an old house – I was deeply intrigued and curious about where such music could come from. It's an image I still can vividly remember.

Contributor: Seth Nehil
Age: 31
Location of memory: Bedroom, Hillcrest St., Kalamazoo, MI
Year of memory: c. 1987

—

None specific

When I was a young child, just beginning to walk, our family had a mid-40s vintage Bakelite radio that sat on top of a bookcase in the bedroom of our apartment. There was no television in those days, and I don't even think we had a record player, so the radio was on a good deal of the time with music, variety shows, dramatic programs, etc. I remember being curious about the sounds, so one day I climbed up the bookshelf and looked behind the radio, hoping to find the "little people" who lived there.

Contributor: Richard Zvonar
Age: 58
Location of memory: Union, New Jersey in the bedroom of our apartment
Year of memory: circa 1948

—

Le Boeuf Sur Le Toit, (The Ox on the Roof), op58, Darius Milhaud

I've given this quite a bit of thought and realised that most of the memories that I have which have been marked by music are in fact memories associated with music which has changed my listening. My memory is from my early teenage years in a small city in Greece, where I grew up. At that time in the late 80s there were no interesting radio stations. The only one which broadcast interesting music was a station in Athens which I could not tune into until after midnight, when radio-taxis and interferences were not as bad as during the day. I would almost always stay up until 3 in the morning, after finishing my homework, trying to tune in to this station. They played what I then called 'non-music', but what I now know is modernist and avant-garde. One night they broadcast a piece played by an orchestra with an infectious Latin American-sounding melody which kept being interrupted by another orchestra playing something else causing a lot of cacophony. I kept moving the antenna, changing the position of the radio around the room, moving my body – nothing. I then realised that it was impossible for the sound of another orchestra to interfere – the interference I usually heard was taxi-drivers speaking! The piece was indeed for two orchestras playing two completely different things but occasionally tuning into one-another. It was a celebration of melody and noise. Something had happened to my listening to music that night. I never moved my radio again when I heard interferences – and rather than feeling frustrated I felt privileged that I was in a way involved in composing the music I heard; my listening depended on my place within the city, on where I had positioned my radio, on how many taxi-drivers worked the night I decided to turn my radio on.

It was not until nearly 10 years later that I heard the same piece by chance on Radio3 in London and went out to buy it. Every time I hear it I remember that strange night in my room.

Contributor: Mikhail Karikis
Age: 29

Location of memory: Thessaloniki, Greece, my room
Year of memory: one night in the late 1980s

—

n/a

When I was 13, I joined the Radio Society of Great Britain and bought a short wave radio receiver [Lafayette HA-800], with a view to passing the exam necessary to become a Radio Ham. Although this never happened, hearing the fuzz and static through the headphones radically influenced the way I listened to sound. In addition, there was also the cultural experience of accidentally tuning into Radio Bulgaria or hearing amateur radio buffs transmitting from even further afield. The antenna was merely a long piece of speaker wire attached to the chimney on the roof. But realising that the air was swamped with radio waves was a revelation.

Contributor: Michael Harding
Age: 47
Location of memory: A farm in southern England
Year of memory: 1971

—

I STOLE YOUR LOVE, Kiss

TO ME, RADIO HAS THIS AMAZING ABILITY TO OPEN PATHWAYS AT DIFFERENT POINTS IN YOUR LIFE. WHEN I WAS JUST ABOUT TO HEAD INTO HIGH SCHOOL A FRIEND OF MINE AND I WERE LISTENING TO A LOCAL RADIO STATION 4ZZZ, TO THEIR METAL SHOW TO BE PRECISE. WE HEARD A BAND THAT IMPACTED ON MY MUSICAL LISTENING FOR A GREAT DEAL OF MY TEEN YEARS. WE'D ONLY CAUGHT THE END OF

THE TRACK, BUT WERE SO STRUCK BY THE SOUND WE WAITED PA-
TIENTLY FOR THE BACK ANNOUNCE AND THEN STARTED A MISSION
TO FIND MORE OF THEIR RECORDINGS. A WEEK OR SO LATER THAT
SAME FRIEND DROPPED OVER WITH A CASSETTE FEATURING FOUR
VERY DOLLED UP GUYS SURROUNDED BY LOADS OF WOMEN ADORNED
IN LEATHER AND LACE – THEY WERE THE MUSICIANS BEHIND THE
TRACK WE'D HEARD THE WEEK BEFORE – KISS. THE MEN FEATURED
DISTINCTIVE MAKE-UP, AS DID THEIR FEMALE ONLOOKERS, AND
WHILE I FOUND THIS SOMEWHAT NARCISISTIC (NOT THAT I KNEW
THAT TERM THEN), I COULDN'T HELP BUT BE CURIOUS ABOUT WHAT
KIND OF BAND WOULD WANT TO, IN THE TERMINOLOGY OF THE TIME
– 'PERVE' ON THEMSELVES. ANYWAY, WE POPPED ON THE CASSETTE – I
HAVE CLEAR MEMORIES OF BEING TAKEN ABACK BY THE QUALITY OF
THE RECORDING – IT SOUNDED FAR WORSE THAN WHAT WE'D HEARD
ON THE RADIO THE WEEK BEFORE – MY FRIEND'S BROTHER HAD BEEN
LISTENING TO THE TAPE SINCE '77 I NOW SUSPECT. THE GUITARS
RANG OUT, THE FIRST FEW NOTES SLIGHTLY OUT OF KEY A RESULT
OF THE TAPE BEING CHEWED, YOU GOT THE SENSE SOMETHING BIG
WAS COMING – THIS HAD THE EPIC NATURE OF SO MUCH OF THE MU-
SIC THAT HAD BEEN BOMBARDING ME AND MY SCHOOL ASSOCIATES –
GUNS'N'ROSES, LED ZEPPLIN AND SO MANY OTHER ROCK BANDS HAD
COME TO LIGHT IN THE SCHOOL YARD. I'M NOT SURE IF IT'S THE FACT
I WAS ABOUT TO HIT THE WONDERS OF PUBERTY AND MY MIND WAS
WONDERING, OR MORE PAUL STANLEY'S SHRILL VOICE, OR THAT FACT
THAT I'D BEEN LISTENING TOO MUCH TO SLEAZE METAL ACT FASTER
PUSSYCAT AT THE TIME, BUT ONE LYRIC REALLY CAUGHT MY ATTEN-
TION IN THE SONG. A LINE CHANTED TO ME BY MY FRIEND WHEN THE
APPROPRIATE MOMENT ARRIVED: 'I WANT TO FEEL RIGHT DOWN YOUR
PANTS AND YOUR TITS'

WOW, MY FRIEND WASN'T JOKING WHEN HE SAID THESE GUYS WERE
FOR REAL. TO AN IMPRESSIONABLE MIND HERE WAS A BAND THAT

REALLY DIDN'T TAKE PRISONERS. THEY WEREN'T GOING TO CLOAK THEIR LYRICS IN INNUENDO, NO THESE GUYS WERE GOING ALL OUT TO SHOCK AND TO ROCK, THERE WERE NO TWO WAYS AROUND IT. I JUST COULDN'T BELIEVE MY EARS. THIS LYRIC STUCK WITH ME FOR THE NEXT COUPLE OF YEARS. EVERY TIME I HEARD THAT SONG, I WAITED FOR THE LINE THAT HAD SHOCKED ME THAT FIRST LISTEN. THEN ONE DAY A LITTLE WHILE LATER I BOUGHT A COPY OF THIS AL-BUM ON CD, I POPPED IT ON AGAIN, HAVING NOT LISTENED TO IT FOR SOME TIME. APART FROM THE CLARITY OF THE OPENING GUITARS, AND THE FACT THAT THE LAYERS OF DUST THAT HAD MADE MY OLD COPY OF THE CASSETTE LIKE LISTENING TO SOUND TRAPPED UNDER A THICK HAZE OF AUDIO SMOKE, THE LYRICS WERE SUDDENLY SO MUCH CLEARER – IT WAS LIKE LISTENING TO A NEW BAND. THEN IT CAME TIME FOR THAT INFAMOUS LYRIC ...'HOW DOES IT FEEL WHEN YOU FIND OUT YOUR FAILIN' YOUR TEST'...

DID I JUST HEAR THAT RIGHT? I SKIPPED BACK OVER THE CD TRANS-FORMING IT INTO GLITCHING WRECK (GOD BLES THOSE OLD CD PLAY-ERS) AND LISTENED AGAIN ...'HOW DOES IT FEEL WHEN YOU FIND OUT YOUR FAILIN' YOUR TEST'...

I JUST COULDN'T BELIEVE, I WENT BACK AGAIN. AND AGAIN. I PULLED OUT THE CASSETTE AND LISTENED – SURE, IT COULD BE ...'HOW DOES IT FEEL WHEN YOU FIND OUT YOUR FAILIN' YOUR TEST'...

BUT I JUST COULDN'T BELIEVE I WAS SO FAR WRONG WITH THOSE LYRICS OR IN FACT THAT I'D NEVER PAID MORE ATTENTION AFTER THAT FIRST LISTENING EXPERIENCE WITH MY OVER ENTHUSIASTIC FRIEND.

I ENDED UP CONSULTING ANOTHER FRIEND'S LYRIC SHEETS FOR THE ALBUM, IT WAS TRUE, SOMEHOW UNDER THE WASH OF STATIC AND

HISS I'D BEEN HEARING ONE OF THE DEFINITELY LINES FOR KISS
(IN MY BOOKS ANYWAY), A LINE THAT NEVER EXISTED BUT CAME TO
SUMMURISE EVERYTHING THAT KISS HAD MEANT TO ME – NO HOLDS
BARRED, UNABRIDGED COCK ROCK. TO THIS DAY, I CAN'T HEAR THAT
SONG AND NOT THINK ABOUT THE GAP BETWEEN REALITY AND BE-
LIEF – A THIN LAYER OF DIRTY MAGNETIC TAPE, SPARKED OFF BY A
BRIEF MOMENT OF RADIO.

Contributor: LAWRENCE ENGLISH
Age: 28
Location of memory: BRISBANE
Year of memory: 1989

—

entirely unknown

I have a great deal of memories associated with radio because radio has such an
influence on my musical and sound related interests. In this case my memory is
not so much connected to a particular song or artist but to the understanding
of the nature of radio itself. Until a certain point I always understood radio to
be a local phenomenon, that is, the broadcasts were from the area where I was
living (upstate New York at the time). Radio provided entertainment for people
and there were a few options either with types of music or news and talk. I used
to scan the analogue dial to explore the different signals coming in and for some
reason I always felt there was something else "out there", an other world in the
imagined space of the ether, frequencies you could detect but not clearly tune
in to. I guess my father saw my curiosity because one day he brought me the
strange radio from the basement and said I might be interested to play around
with it. Indeed I did. This radio was different. It had more bands to search,
not just the usual AM (amplitude modulation) and FM (frequency modulation).
This one also had SW1, SW2, MB and LW. The abbreviations were written out

too; Short Wave, Marine Band, Long Wave, which helped arouse one's curiosity. The first exploratory sessions were amazing. Unusual voices streamed in from different parts of the world in unusual programs (more like radio dramas). It was probably my first significant exposure to hearing languages other than English. At once the my world of upstate New York opened up to the world, in it's unbound, unfathomable distances, brought to me via new radio territories. At once I felt connected, "tuned in", in the air of the radio ether or "other" that I previously felt. The unusual noises in between the voices heightened the sensation. They were the evidence of the frequencies swirling in the air, the sine tones, beeps, pulses and drifting static all gave depth to radio. At once I began my studies of the spectrum to listen to the (often temporal, drifting) signals that crossed my radio device. From that point on I preferred the multi-band model over the standard two band one. The most notable memory from this experience is associated with the mysterious Marine Band. Marine Band had very little human content either from my lack of understanding or suitable position in which to catch the signals. But one sound stood out clear and defined from the other noises. It was a continuous monotone pulse that came in an arythmic pattern and seemed to go on endlessly with a few occasional pauses. The patterns would sometimes repeat and sometimes vary. I summized this was Morse Code because we had learned about it in school, that this signaling system was a precursor to the telephone because it was carried long distances over land using wires. It was coded and decoded by a person at each end. I imagined the code was transferring messages from ships at sea among each other or back to land yet one can never really know. But here was Morse Code, alive and still in use, yet in a different medium from what I learned. It was an awakening that radio could be used for many different purposes by many different people. It was no longer an established program produced by a few people, it was a medium for all types of communication, available to nearly everyone to listen to with access to a fairly common "transistor" box (although I already knew that most radios did not have the "special" bands that gave access to the exotic stuff). From then on I continued to explore the outer regions of the radio spectrum in order to temporarily reach new sonic and cultural territories of the world. This early awakening to the possibilities of radio helped influence my current feeling that radio must be considered much more a public utility for broadcast communications rather than the profit driven commercial industry that it primarily is today.

Contributor: John Grzinich
Age: 34
Location of memory: Bedroom, Hyde Park, New York
Year of memory: circa 1980

—

Lay Down Sally, Eric Clapton

As insignificant as it sounds, I remember very little from when I was really young, so this one Alabama summer memory is special. If I had to guess, it is one of the first times I remember being left in the car alone while my father ran an errand. I remember riding in the passengers seat of my dad's chocolate and rust datsun as he cut through the K-Mart parking lot, weaving in and out of parked cars and shopping carts. When he reached the Krispy Kreme doughnut shop he promised me he would be right back and left the car running, with the am radio still playing. The shop had large windows, so he was always in clear site, plus it was the late 70s/early 80s (I place this memory around my 5th or 6th summer, which either would have been 1979 or 1980) and I don't think he was too concerned about someone driving off with the datsun and me while he was buying doughnuts. I remember having a new feeling of being alone for the first time, as we were a family of 5 in a small house and I was always with someone at all times up until that point. It didn't take long for my dad to return with a boxed dozen of glazed – but it seemed like an eternity and the line to pay moved in slow motion while the vinyl seat was sticking to my leg and the song in question crackled out of the tinny speaker.

Contributor: Craig Ceravolo

Age: 30
Location of memory: Midfield, Alabama – a western suburb of Birmingham
Year of memory: 1979 or 1980

—

Take Five, Dave Brubeck Quartet featuring Paul Desmond

Sitting at a table in a noisy restaurant listening to my friend Phil and his parents talking, I heard way in the deep acoustic background something familiar in the muzak (I'm pretty sure the owners were piping in a radio broadcast). I could only detect the most salient features of the spectral high end. After focusing on it or a minute or so despite all the noise I finally recognized it as "Take Five," a jazz tune I'd listened to countless times as a teenager. This experience made me realize how the specific features of that recording were deeply engrained in my memory and consciousness and I could recognize those features through almost any degree of noise or distraction.

Contributor: John Bischoff
Age: 54
Location of memory: in a Chinese restaurant in San Francisco having dinner with my friend Phil Harmonic and his visiting parents
Year of memory: ca. 1980

—

Not sure. Elvis. Any early Elvis will do.

When I was a kid (from age 7 on until 13 or so) I had inflammation of the middle ear at least twice a year. A painful experience which was relieved by the fact that I was allowed to stay on the couch in the kitchen – times or moments when I felt close and somehow on close terms with my mother, who usually hardly allowed intimacy. I had a small blue transistor radio (FM) with a poor sound. For some reason (guess I tried to find out how to make it sound better) I started to dismantle the thing and soon had it opened with the speaker laying bare. Without the resonance body the thing sounded even poorer than before – nevertheless I can well remember the change it did to some Elvis song, that became high pitched and metallic – somehow more real than as it had been inside the box.

Contributor: Achim Wollscheid
Age: 45
Location of memory: Trier, Germany, Bergstrasse 66, the kitchen
Year of memory: Around 1967

TRANSMISSION 5: SONOROUS LANDSCAPES/ CULTURAL VIEWPOINTS

Being struck by the coming into being of sympathetic resonance: between place and audition, location and its audible life. What planetary reverberations spread forth from this unfamiliar voice? What aerial bodies find their place at this spot, on this ground, to take root and bloom into hybrid cultures? Sidestepping, aiming for other routes, captured by vital noise that will never depart this place, that will remain a landmark of musicality to which listening returns.

Le Freak, Chic

Vacation, The Go Go's

The Magnificent Seven, The Clash

Blood, This Mortal Coil

Phantasmagorica, The Damned

Brazil Classics Vol 1

Le Marquises, Jacques Brel

Flutes of Rajasthan, Karna Ram Bhil

I'm Like a Bird, Nelly Furtado

call to prayer at dawn, Jordan

I'm gonna lock my heart and throw away the key, Billie Holiday

Hong Kong Garden, Siouxsie and The Banshees

This Is the New Shit, Marilyn Manson

Comfortably Numb, Pink Floyd / Scissor Sisters

Heart And Soul, T'Pau

Soul City, The Partland Brothers

I'm beginning to see the light, Tommy Dorsey

Le Freak, Chic / *Vacation*, The Go Go's

My fondest memories of music often take place when I am travelling – when my emotions and senses are at an ultimate high. So, here are memories at various times of my life (location and year will be outlined in each one): – First trip with my parents who brought my sister and me to Disney World in Orlando, Florida. I remember being dragged amongst the crowd, sniffing cotton candy, being bombarded by a giant Goofy and Minnie Mouse, and hearing on the loud speakers "Freak" by Chic. 1978. I still love that song.

Travelling to Montreal, Quebec in my parents car. My sister had bought her first walkman. The first song that came on the radio was "Vacation" by The Go-Go's. That's when I developed a fantasy of being in an all-girls band and wanting to learn to play guitar. 1982.

Contributor: Nadia Bartolini
Age: 30

—

Magnificent Seven, The Clash

It was the summer of 1981 and I was doing time at an office supply warehouse on West 18th street. My workmate and I used to tune in the radio to WBAI's "Stormy Monday" on the sly between the stacks of cartons of Hammermill copier paper. But also – dare I admit it? – Scott Muni's "Things from England" just so we could seethe and complain that WNEW and Muni were basically over-the-hill sell-outs who ignored everything that was actually happening in England and everywhere else. How could Muni play the latest Genesis while totally ignoring Wire, Gang of Four, Blurt, the Pop Group, the Slits, or the Fall ... and their delirious sound of things falling apart. [It can be said that this revelation – the abysmal emptiness of commercial FM – led to my introduction to WFMU in 1981 and some years later, in 1986, I joined WFMU as a DJ to fight the good fight for "our" right to serenade our own nervous systems with our own sounds and lead sonorous lives.] It was a hazy, humid June day and I was walking languidly west on 14th Street, returning from "lunch" with my Indonesian girl friend. I say quote lunch unquote because we didn't eat anything – just stared and sighed at one another sitting on the back of a park bench in Stuyvesant Square Park along Second Avenue. As I walked back to work, I thought only of her – so young, so devastatingly beautiful and petite, so breathtakingly silent, tranquil, self-assured, intelligent that whenever I was around her I would crack and end up in the middle of some Three Stooges skit – goofball, in other words. This phenomenon of serenity combined with beauty can be so disarming as to make you physically sick – giddy and nauseous … and goofball. I thought of how I – if we weren't meant to be forever – would write her into my first novel, if that was ever going to happen. I had just passed the Palladium [now gone], conniving how I could get myself fired and collect unemployment.

As I crossed Broadway, I suddenly heard "The Magnificent Seven" by the Clash on a radio in a car turning right, then I heard it on a radio turned up loud in the strange name-brand electronics store, and then the bad luggage store, the discount mothballed fabrics store, the bright red soda deli, the past-due-date pharmaceuticals discounter, other cruising open-windowed cars, a guy with a big sound system on his shoulders, everyone in the world [as I was experiencing it, anyway] was playing that song, with people singing along, bobbing their heads, waling with various expressive bounces to the funk-driven beats and rap-oid lyrics: "Don't you ever stop long enough to start? / To get your car outta that gear / Karl Marx and Friedrich Engels / Came to the checkout at the 7-11 / Marx was skint, but he had sense / Engels lent him the necessary pence…" This led to extremely warm whimsically lofty feelings that can only be described as Whitman-esque – Carl Sandburg probably also wrote about it – this epiphany of momentary fellow-hood and the mystifying significance of sudden synchronous instant – every radio everywhere seemed to be tuned to the same song on the same radio station, all singing along to the same lyrics "Better work hard – I seen the price / Never mind that it's time for the bus / We got to work – an' you're one of us / Clocks go slow in a place of work / Minutes drag and the hours jerk..." Everyone experiencing the same instant of joyous recognition in the beats, the lyrics, but also the fact that everyone knew the tune and had already been living the words for years. And so for that one instant, that one skinny sliver of time you could nod to a Puerto Rican, a Dominican, a Jamaican, an Indian, a disco maniac, someone from the Bronx and they would nod back that nod, that important identification-verification twitch, in passing and no one thought you were homo, weird, addicted, hitting them up for loose change, about to rob them. Soundtrack of the human working world. And then I did something I NEVER-EVER do, I began singing along at the top of my lungs to the music that seemed to be emerging from every open window everywhere – "Wave bub-bye to the boss / It's our profit, it's his loss / But anyway lunch bells ring / Take one hour and do your thanng!" – from Broadway over to 6th Avenue.

This moment of pure camaraderie through the portal of sound would probably be shot in slo-mo for the film version. It would be the document of a glorious feeling I have never felt since. It is the last time the center embraced the margins and punk sounded like Woody Guthrie in the Summer of Love, when pure musical joy became an issue of class consciousness, when weirdness and straightness merged as part of the same insane work drudge personality continuum, when all first, second, third, and fourth world peoples for 5 minutes spoke the same language. And now that I think about it – and not at all sentimentalize – there have only ever been a few moments during my life that this could have happened and ever DID happen – "This Land is Your Land," the night John Lennon was shot, Sly & the Family Stone, James Brown, the Beatles, "Wordy Rappinghood," something by Curtis Mayfield or Marvin Gaye or Marley, I don't know... Hope and clarity, satori and epiphany... but, however memorable, what is five friggin' minutes in the past quarter century?! "News flash: vacuum cleaner sucks up budgie / Oooohh...bub-bye..."

Contributor: Bart Plantenga
Age: 50

Location of memory: New York City, 14th Street
Year of memory: 1981

—

Blood album, This Mortal Coil / *Phantasmagorica* album, The Damned / *Brazil Classics vol. 1*

i made a yearly trip from san francisco to spokane wa. to visit friends. i taped music to play on the 3 day drive. whenever i hear this music, it transports me back to the drive through northern california, oregon, and washington.

Contributor: Deborah Valentine
Age: 49
Location of memory: driving north from san francisco on hwy 5 through oregon and washington
Year of memory: 1992

—

Les marquises, Jacques Brel

During all my childhood and beyond, I never had a space dedicated to a musical listening. This period is however inhabited by a regular musical feeling. The radio was everywhere... top of the refrigerator of my kitchen, of the bar, without forgetting in the car. The songs of this time thus come to infuse in these spaces. Rarely in the foreground, they paper the bottom of the conversations. I remember these songs coming to be crushed in daily sound flux (of the pressure cooker before the lunch, of the smoked out hubbub of after match of football).

Contributor: Eric La Casa
Age: 36

Location of memory: In the howling chaos of sound flux, this song seemed to me a lull. As a light breeze which suddenly invades me and opens another prospect to me. This voice, the violins... something transported me. I do not have any more precise memory of the place but imagine this song in the morning of a kitchen, the window open on a garden, and this fog of my native area...
Year of memory: 1977

—

Flûtes du Rajasthan, Le Chant du Monde/Karna Ram Bhil playing the nar

While living in Andalucia, Spain, in a small "cortijo" (little farm house) in the Sierra Nevada mountains, with no T.V., and very few modern conveniences, I listened to the radio often, every day. At the time I was working on my Doctoral Disseration on the study of singing while playing the flute. I was trying to gather as much information as possible as to where around the world people blow into tubes and sing simultaneously. Meanwhile I was writing a manual on how to teach flutists to do the same. One day, in the evening, I heard quite to my surprise the most intriguing, fascinating, beautiful and yet strange music I'd heard yet: The sheperds from Rajasthan, India, playing the nar (similar to the nay) and singing in the most glutteral voice I'd heard since the Tuvan singers. I flipped out: even being in the "middle of nowhere" I could still do my research by radio.

Contributor: Jane Rigler
Age: 38
Location of memory: Kilometer 3 del Pantano, Quéntar (Granada), Spain
Year of memory: 1995 or 1996 (not sure)

—

I'm Like a Bird, Nelly Furtado

This song was playing on video hits when I first heard it and thought it was light pop trash, with a slightly strange video clip where a girl falls out of a tree. After that I started hearing it on the radio everywhere; driving in a friend's car, out walking past shops and bars, and in the kitchen of my house when I was doing dishes – living with a flatmate who had the highest possible taste in literature and film, but loved top 10 pop music (Ronan Keating, Faith Hill), which started as an affront to my experimental electronica aesthetic but was actually kind of fun to dance to while doing the dishes... At certain moments in life, particular things resonate for you. I had just been made redundant from a dreadful and boring office job, and was taking the first steps towards making a plan to realise my dream of travelling the world and recording the sound of bridge cables. Every time I heard this song, it reminded me that there really was nothing holding me to my life in Melbourne, and I actually could pack up and fly away. Something about the melody seemed to encapsulate that impulse, and even now as I write this listening to the single which I recently bought for $1 in a second hand shop (they don't play it on the radio so much anymore), it lifts my spirits. "I'm like a bird, I wanna fly away, I don't know where my soul is, I don't know where my home is." Soon after this I moved to Sydney for 3 months, got the bridge recording trip financed, and boarded the plane to Ho Chi Minh City. A couple of months later, after recording bridges in the Mekong Delta, Rotterdam, Helsinki, Berlin and London, I found myself in a fabulously kitsch circular bar (unchanged since the early 60's) on top of the Novy Most (Bridge) over the Danube in Bratislava. Drinking my grape cider while looking out over the 14th old town on one bank of the river, and monstrous clusters or identical Russian apartments on the other, this song came on the radio. It was a classic euro-trash pop station, strangely out of place in the surroundings, yet perfectly suited to the only other customers, a family with small children. That moment of hearing this particular song again gave me a sense of how far I had come on this journey, and – not completion – but something about the circularity of life, and how even though they appear to be separated by vast distances and the passing of time, these moments where our dreams intersect with our realities are completely interconnected.

Contributor: Jodi Rose
Age: 34
Location of memory: Melbourne, Australia various radio spaces and Bratislava, Republic of Slovakia in a bar on a bridge
Year of memory: 2000 – 2002

—

call to prayer at dawn from a minaret in Jordan.

Sitting on a rock at dawn, wrapped in a blanket, watching the sun rise over a labyrinth of caves, listening to the microtonal call to prayer echoing over the rocky landscape from the Bedouin town behind Petra. Due to the invasion of the tourist industry (of which I was of course a part), the Bedouins that previously lived in some of the caves have been evicted and pushed further and further back from the public 'front' of Petra, at the same time as being 'contained' in purpose-built villages. Every evening, the extensive network of caves (at least 25km2) are patrolled by police who clear out both lingering tourists and Bedouins. I had befriended some Bedouins, with the aim of making recordings on a cumbersome reel-to-reel tape-machine loaned to me by the University of Cambridge's ethnomusicology department. In the event, I left the tape-machine at my hostel as it was so unhandy and unsuited to the task of basically hiding out in these caves overnight. But the experience of that sound piercing through the rocks together with the day's first sunrays has stayed with me. I don't know exactly if you can call this a 'song' – for a Muslim it wouldn't count as one, I guess. And as for 'public broadcast system' – well, I guess that these days a lot of those calls-to-prayer are recorded, so maybe it is mediatised after all (??). Anyway, it's one of the clearest sound-event memories I have.

Contributor: Juliana Hodkinson
Age: 33
Location of memory: Petra, Jordan
Year of memory: 1992

—

I'm gonna lock my heart and throw away the key, Billie Holiday

I vaguely remember hearing Billie Holiday for the first time on the radio. I was
in high school, a friend of mine told me that I would like Billie Holiday shortly
before that. When I heard her for the first time, I thought that her rhythm was
loose, the voice was expressive, but not particularly beautiful. I soon became
addicted though and until about 1988 I listened to a lot of Billie Holiday. Those
were the days when I would tape thing off the radio. I still have some of those
tapes. Although this happened in the very early 80s, it was still the 70s in a
way, which meant that there was still a trace of 60s culture in and around San
Francisco where I grew up. Radio was a part of that in a big way (as was guer-
rilla TV on channel 26 during late night–there was a show called Video West
which showed very psychedelic short films; political stuff like all kinds of non-
mainstream sexuality, dykes on bikes, gay cultural things, I remember there
was something on geriatric sex once).
As kids, we thought a lot about what the 80s would be like, because we were
desperate for it to be different from the 70s. Well, some things got better, some
things got a lot worse. In a way, the texture of the memory doesn't lend itself to
singular artist/group–for me it was more about the kind of college or lefty stations on the left side of the dial. I listened to a lot of punk, "new wave" (though
I was told shortly after that new wave didn't exist because only failed or fake
punks were new wave); a lot of stuff from the 20s, 30s...

Contributor: Simon Leung
Age: 40

Location of memory: San Jose, California
Year of memory: circa 1981

—

Hong Kong Garden, Siouxsie and The Banshees / *Comfortably Numb*, Pink Floyd &
Scissor Sisters / *This is the New Shit*, Marilyn Manson

Recently, British DJ John Peel died, unexpectedly (more unexpectedly than
usual, that is). Listening to a radio call-in, men of about my age would tell of
their teenage experiences of listening to John Peel under the bed clothes, capti-
vated by the music, his dry enthusiasm, and the sense that something "big" was
happening to the music scene. I was lucky: at 15, I got a radio combined with a
tape player for my birthday. I didn't have headphones, so I would press my ear
to the speaker (somewhat painfully, as the plastic shell was dimpled) and keep
the sound low. Evidently, just like my peer group, it was Peel we waited for ~
under the blankets, towards midnight. (No wonder we were all knackered at
school the following morning.) Towards the end of 1976, and into 1977, although
I didn't know explicitly know or understand it, something was happening to
music: a new wave. The press would call it punk: a worthless musical form
played by worthless people. In particular, my dimpled ear remembers the first
chimes of Siouxsie and the Banshees' remarkable Hong Kong Garden (their first
single: I am proud possessor of a limited edition gatefold version, only a 1,000
ever made they said). By the end of 1977, we were all selling our Genesis (etc.)
LPs. Except I never sold my Pink Floyd LPs.

Recently, I was in a clothes shop, when a dreadful dirge came on the in-store
piped music. It sounded like a collision between the Bee Gees, New York disco
circa 1974, and a children's band: clumsy, woeful, irritating enough to catch
the ear and not let go. About half way through, I realised I recognised the song
~ with absolute horror, it was clear someone was murdering Pink Floyd's Comfortably Numb. Who was slaughtering the song, and so sacrilegiously? Shop

assistant had no idea. But the song was everywhere, and it turned out to be the Scissor Sisters. Now I understand that this is symptomatic of what is happening to music in an iPodded ~ iPodized ~ world. 10,000 plus songs on one tiny machine erase the differences between songs, turn them into 1974 New York disco fodder. A vast musical phantasmagoria in which songs become ghosts wandering through a musical landscape incapable of hearing the difference. Why not a disco version of Comfortably Numb?

As Marilyn Manson might say, This is the New Shit. Can't wait for the Scissor Sister's version of that. Out with the new, in with the old. But this is not a complaint about the meaninglessness of contemporary music, rather a simple recognition of the musical phantasmagoria that pervades contemporary life. Not so much a soundtrack of our lives, complete with pained and excited ears, as a pervasive iPodded background music that will stream endlessly, almost unrecognizably, past us: forever set to random, repeat all.
Hope I get an iPod for Christmas.

Contributor: Steve Pile
Age: 43
Location of memory: Barton St David, Somerset, UK; and London, UK.
Year of memory: 1977/1978 and 2004

—

Soul City, The Partland Brothers / *Heart & Soul*, T'Pau

My first overseas trip...I'm flying to see relatives in Italy...I'm 15 years old. I've never spent so much time flying over such a large mass of water before. I'm a bit nervous, cause I've seen all those disaster flight movies like Airport 77 so I'm just waiting for this sucker of a Boeing to take a nose-dive at any moment, where half will die on impact and half will be eaten alive by freakish Jaws-clones. For a buck, the stewardess (politically correctness wasn't hip yet back then) handed me a cheapie headset to plug into the armrest. To my delight, I find that not only can you watch (and listen) to educational films such as the Muppets Take Manhattan but there are also many music stations to choose from by turning the dial....one is classical, one sounds pretty ethnic to me, the other is Barry Manilow-city...yawn, yawn, yawn...nothing too interesting to a 15-year old...then...oh joy! I stumble upon a station and immediately recognize the energetic, spasmic sound of (what is now referred to) Retro 80s music. The selection of tunes (set on 'Repeat') was meager but how relieving it was to hear the familiar sounds of what I loved so dear...the flight suddenly didn't seem to doomed anymore.

That day, I developed a particular fondness for two specific songs and even today, some 16 years later, whenever I hear those two songs, I am immediately transported back to that flight over the Atlantic...full of fear, exciting and wonder of all the things I'm about to discover in this far-away country...

Contributor: Tina Bartolini
Age: 32
Location of memory: Somewhere over the Atlantic Ocean!!
Year of memory: 1987

—

I'm beginning to see the light, Tommy Dorsey

When I was a little girl, approximately in 1945, and came home from school my mother sat by her handloom. To make place for the handloom we had moved the kitchen table aside. I had borrowed a book from the school library, the "Jungle Book" by Rudyard Kipling. I sat by the kitchen window and read in the book about Mowgli and the bandarlog monkeys.
Suddenly I heard beautiful music from our radio. It was a bigband playing "I'm beginning to see the light". When I think back I'm almost sure it was Tommy

Dorsey. I was humming and reading at the same time and the music was so wonderful (and the book too). I think I was around 11 years old.

Contributor: Karin Nilsson
Age: 70
Location of memory: Skuga, Värmland, Sweden
Year of memory: 1945

TRANSMISSION 6: INTRUSIONS & REVELATIONS

The frame is always brushed by outside influence, the content finds its antithetical intruder in the form of formless tappings, knocks on the door of attention: vessels struck by other fingers, surfaces touched by unfamiliar hands, bruised by exterior forces. Strangers in a strange land, where X is the perennial variable supplying voluptuous antagonisms to the haven of appearance–X-ing out, making over, licking the ear, annoying audition...

Song to the Siren, Tim Buckley

Free Fallin', Tom Petty & the Heartbreakers

Dog Breath, In the Year of the Plague, Frank Zappa / The Mothers

Holiday In Cambodia, Dead Kennedys

Mon Ideal, Ti Paris

Goody Two Shoes, Adam and the Ants

Lay Lady Lay, Bob Dylan

Toccata & Fugue in D Minor, J. S. Bach

Holiday In Cambodia, Dead Kennedys

4'33", John Cage

Song to the Siren, Tim Buckley

I HAD WRITTEN INTO A RADIO PROGRAMME. I WAS ABOUT 17 YEARS
OLD. I HAD DEFACED A POSTCARD IN A SURREALISTIC MANNER TO
GRAB THE ATTENTION OF THE DJ. YOU NEVER EXPECT TO BE PICKED
OUT. BUT ONE NIGHT WHEN LISTENING TO HIS SHOW, HE SAID MY
NAME AND THAT I HAD REQUESTED THIS SONG. I HAD TILL THAT
POINT ONLY HEARD A COVER VERSION OF IT, AND WAS DYING TO HEAR
THE ORIGINAL. HE PLAYED IT, I TAPED IT ONTO A C90 TAPE, AND IT RE-
MAINS ONE OF MY FAVOURITE SONGS BECAUSE I FELT EXTERNAL TO
MYSELF IN THE SHAPE OF A SONG. AND BECAUSE I FELT THAT SOME-
ONE ELSE FELT THE SAME WAY I DID ABOUT OBSCURE RECORDS THAT
NO ONE ELSE I KNEW KNEW ANYTHING ABOUT.

Contributor: Shumon Basar
Age: 30
Location of memory: My bedroom, Blackpool England
Year of memory: 1992

—

Free Fallin', Tom Petty & the Heartbreakers

My friends and I went to an abandoned missile silo to record some music in the
underground rooms. Whether it was night or day, flashlights were needed to
see in the rusted, asbestos ridden underground tunnels. After exploring the
launch silos (which were 165 feet deep and filled with water) we walked to a
huge dome area that was sectioned off into rooms. Walking up stairs, I noticed
an old bathroom at the top. Distracted by my find, I turned my flashlight to the
right where a room was. I failed to notice that there was a gaping hole a step
in front of me, walking forward and directly into the hole. I fell 25 ft. into the
darkness. The fall seemed to take forever while my life flashed before my eyes.

I landed on a gigantic vent cover that layed on a large pile of scrap metal which padded my impact. A shard of metal knocked my hat off and came millimetres from entering my skull. My friends descended the stairs and found me on the level below laying there dazed and quite dirty from all the soot in the silo. They then helped me out of the silo and to the car where we drove to the emergency room. After a confusing, multiple physician examination, an RN wheeled me on a bed into a cold, dark x-ray room. Lying there alone while he prepared for the tests elsewhere, I stared up at the ceiling and noticed that the ceiling tiles had been removed for a maintenance. Exposed was concrete, wire, and old pipes that reminded me of the dark, cold silos. A radio was playing on the other side room as I lay there and waited for the x-rays to begin. A song came on, starting with a simple guitar riff as Tom Petty's voice stared to sing "She's a good girl, loves her mama, loves jesus and america too, She's a good girl, crazy 'bout Elvis, Loves horses and her boyfriend too". It took a couple of moments to realize, but as he began to sing "And I'm free, free fallin, Yeah I'm free, free fallin," I lost it! Being that I had just fell 25 ft and Tom Petty was singing about free fallin!! I screamed "Turn It Off!! Turn It Off!" It freaked me out! Now, years later I look back and think of how ironic it was hearing that right after my accident. I'm still haunted by the incident every time I hear "And I'm free, free fallin, Yeah I'm free, free fallin"

Contributor: Samuel Cooper
Age: 27
Location of memory: Denver Generial Hospital, Devner, CO
Year of memory: 1999

—

Dog Breath, In The Year Of The Plague, Frank Zappa & the Mothers of Invention

I was riding on the backseat of the car with my parents, after a Sunday walk. We were on a long straight street through the forest, heading home, when in the radio a speaker said there was a new band from the States combining Rock, Jazz, Opera singers, and anything else in a strange mix, never heard before. Then they played excerpts from the song, still speaking over it. I was absolutely delighted by this music redefining the little I had heard so far at the age of 6. Revelation! Maybe at this moment I decided to go into music? But I was sad because my parents cancelled the song much too soon, conservative as they were. However, 10 years later I bought the album Uncle Meat by the Mothers of Invention (at that time hard to find in my hometown!), since I got interested in Zappa's older stuff, having heard some at a friend's place – and to my surprise there it was!

Contributor: Marc Behrens
Age: 34
Location of memory: Close to Messel, Germany
Year of memory: Possibly 1976

—

A Holiday in Cambodia, Dead Kennedy's

I grew up in Helsinki Finland and lived there until I was 30. The first independent radio station, Radio City, opened in 1984 as a result of a bunch of kids squatting an abandoned building in Helsinki and turning it into a house for art and live music. I believe I was 19 then. Until then, all radio stations were controlled by the state. This meant we did not have to suffer from commercials on the radio, but also that there was a certain control, however subtle or innocent, over all the content broadcast in Finnish radio stations. This control was moderately anti-Western. For a teenager like me, this meant I could not trust the Finnish radio to bring me the music I wanted to hear and keep myself updated on. I was into disco music, Jackson Five, Donna Summer, Grace Jones, etc. I loved to dance. So, like many other kids, in the wee hours of the night, I listened to Radio Luxembourg. You needed an old radio for that, one that could pick up the long waves of international radio stations. Also, you could only pick up the faint signal during night, when there was more space in the air. Trying desperately to not wake up my parents, I would lie in my bed with my ear leaning right on the loudspeaker. I would listen and sometimes record with my tape recorder straight off the old radio. Every 5 minutes, the signal would become faint, and I would have to tune in again. And here is the unforgettable memory: I must've been 15 years of age. Suddenly, I hear something I have never heard before. It is "A Holiday in Cambodia" by Dead Kennedy's. I was in awe of the temper of the sound and the lyrics. Without exaggeration: My life changed in that instant. From being merely a good student in school – a glittery disco dancer, I became personally engaged and acerbic in regards to history and politics. I started listening to a whole range of different music from Punk to Bob Dylan. I started hanging out with a different crowd. And some months later, a short article I wrote on the genocide in Cambodia was published in the local newspaper.

Contributor: Pia Lindman
Age: 39
Location of memory: Helsinki, Finland, my own bed room in my childhood house
Year of memory: 1979-1980

—

Mon Ideal, Ti Paris

On Saturdays I remember my mother would take her time to clean up our house and cook while she whistles to the different tunes coming from the radio. "Radio Haiti", a radio station that played a variety of artists on Saturdays not being broadcasted like the "mini jazz" bands, the standard "Haitian compa" (Haiti's popular music) would play artists that were not well celebrated and of course Vodou music that was not broadcasted by the "elite" station. The first time I heard Ti Paris I was about eight or nine year old. His melodic style, lyrics and sense of humour (proven to be very political) were totally different from the other musical groups I was accustomed with. As a nine year old, singing a Ti Paris song like "Mon Ideal" would sometime bring some scolding from a parent due to the racy flavor of the lyrics. Basically some Ti Paris poems were not to be sung in the presence of an adult.

"Ti Paris di trop' betise"... Ti Paris curses too much. That is what the adults have always said about his music. The first time I was caught singing "ay Lina cheche yon batwel pou bat' boudam'"..., I was told not to sing that song. Period.

I loved that song and I wanted to play guitar, make my own guitar and be a Ti Paris. I guess that's when musical taste found it's bud. When I discovered that my father actually had Ti Paris et sa Guitar in his records collection, I invited all my friends to come and listen. We knew all the words and guitar licks by heart. Still today Ti Paris et sa Guitar remains one of my favorite records. Ti Paris was only playing a four string guitar that I later learned he'd fabricated himself. With his charismatic voice and bluesy style guitar along with his deep and humorous lyrics (that sometime would land him in jail along with a beating) he used his tongue (clicks of the tongue just like the bushmen in Africa) as a rhythmic pattern to accompany his guitar along with a "tcha-tcha" (a gourde shaker) and a drum. "Mon Ideal", as a fact, the whole album "Ti Paris et sa Guitar" turned out to be my favorite of all time.

Contributor: Karl Jean Petion
Age: 40
Location of memory: My story/memory took place in Port-au-Prince, Ayiti. When our family moves to Petion-Ville, a neighbourhood where the music scene was flourishing
Year of memory: 1972/1973

When I was in second grade, my family moved from rural Pennsylvania to suburban Massachusetts. I had no real familiarity with pop music outside of a few records my parents owned, but when I got to Massachusetts I was introduced to radio rock, initially in the form of Joan Jett & the Blackhearts and the J. Geils Band. I got a little "boom box" which was my favorite item in the world – It had a mono speaker, an AM/FM tuner, and a recording tape deck. I would make compilation tapes from Top 40 radio, as I imagine many kids did. One of my favorite songs was "Goody Two Shoes" by Adam Ant, and I waited vigilantly to get it on tape (asking my parents to buy me the record was beyond my thinking, apparently).

When recording any song off the radio, I was fastidious about starting the tape in the split second between when it was announced and the first note sounded, and stopping the tape the instant the song ended (in those days local DJs were less likely to talk over the beginnings or ends of songs). At the end, of my recording of "Goody Two Shoes," when Adam Ant sang the last "there must be something inside" and the final retro-50s guitar lick was played, I hit stop – forgetting that the song really ended with two snare drum hits. I was devastated – I had missed the final two beats, and the song was incomplete! Then I quickly hit upon a solution: I would add the hits myself. I knew already that I was good at vocal impressions, including for non-vocal sounds. So at the end of my recording, where the snare hits would have gone had I not stopped the tape, I simply sang two hits into the tiny built-in microphone: "Dvvvff, Dvvvff..." I got everything from the onset to the reverb. Confident that I had made a seamless "punch-in" of the final drum track, I rewound and reviewed. To my horror, it didn't sound at all like a hip snare lick – it sounded like a kid spitting into a microphone. I'm good at vocal impressions, and actually put a lot more work into it then I normally think about. I'm a musician and sound artist but have never called myself a vocalist; but most of what I do with sound comes from absorption of my sonic environment, which is closely connected to my ability to emulate. That event may have spurred my interest in creating my own sounds.

Contributor: Michael Bullock
Age: 31
Location of memory: The front stairway of our first house in Marshfield, MA. I
was sitting on the second or third step with my little boom box in my lap, fac-
ing me.
Year of memory: It must have been 1982, since the song had just come out.

—

Lay Lady Lay, Bob Dylan

It was 1989 I was 12, I was walking home from guitar practice in the suburbs
of Philadelphia (exton to be exact) its kind of a yuppie/red neck town. I hated
my guitar teacher because he was forcing me to site read and learn songs by the
who and the police, i wanted to play power chords and just rock out to gnr, the
who was just too "old head" for me. His name was ken he had short hair and he
crossed his legs when he played and I thought he was cheezy. But i went to him
anyway because i could walk from my mom's townhouse through the science
industrial park to the strip mall of suburban shops, where my guitar teacher
had his studio.
On the way home after practice, i was frustrated and pissed, and i stopped to
lite a cigarette. It was getting dark, and it was in the spring, so the light was
really pretty, warmish just past sunset. I heared this sound and this voice, i
couldn't recognize it i believed i had never heard it before, but i could see this
blue light coming out from a loading dock from one of the industrial buildings
down the hill to the left. I thought it was a band playing in there, and i was shy
but really loved the sound, so i slowly walked down there, sort of afraid, because
the workers in the area were always sort of creepy and would heckle me and my
friends when we rode our bikes down there. Anyway, when i finally got close
enough it was just large boom box, in the middle a totally empty garage. There
was no one in site, and it was so beautiful i never wanted it to end. i stayed for
the whole song, but i couldn't make the lyrics out at all, it was too distorted. it

took me a whole year to find out who the singer was and the song. It was Bob Dylan singing Lay Lady Lay from Nashville Skyline, 1969.

Contributor: Madeleine Gallagher
Age: 27

—

Toccata and Fugue in D Minor, Bach

Once upon a time I practice as an electrician. I think I was around 15 years old. We had installed a new organ in the church in my hometown. Everybody who had been working with the organ was invited to the consecration of the new organ. The organist was the very famous blind german organist Helmut Walcha. The first number of the concert was Toccata and Fugue in D minor by Johann Sebastian Bach. Walcha started to play but after a couple of bars he stopped. Then he raised from the chair and called for the church porter. The porter went up the stairs to the organ loft, listened to Walcha and came down again and disappeared for a short moment. Then he came back with a little hammer and went up to the organ loft again. Walcha asked the porter to hold down one of the keys and climbed up on the organ and with the little hammer he adjusted the tune of one of the organ pipes. I could not hear any difference, but Walcha's sensible ears had obviously discovered that the tune of one of the pipes was not correct. How he could hear that I do not understand especially when he plays such a noisy opus as the toccata. After some adjustment Walcha was content and returned to the organ chair and started to play the Toccata again from the beginning. From that moment I love the music by Johann Sebastian Bach.

Contributor: Lennart Nilsson
Location of memory: Sweden

—

Holiday in Cambodia, Dead Kennedys

sometime in 1983/4 i heard Holiday in Cambodia by Dead Kennedys on the famed "cutting edge" radio station 91X. i was staying up a bit late for school the next day, still a good boy, empowered to mine the night air. something about the song moved me, more than the combined efforts of style, technique, skill, and technology. it got me thinking there was more to the cutting edge than post-new wave rock gruel, to feel there was a world outside of my hip suburban enclave where pathos and beauty could co-exist. it was a transcendent sonic assault, beyond the writing, the licks, the chops, names, ideas, it was a revolt of feeling, a kind that rose through my skin. later on i would acquire the album Fresh Fruit for Rotting Vegetables (their first album, featuring Holiday as its closing blow) after first getting hooked on Plastic Surgery Disasters (which was my favorite), and would eventually know the lyrics to all of the DK songs by heart. i would come to despise the whiny hatred of Jello Biafra, the lysergic paranoia of persecution and abuse (which is still valid in occasional doses, frankly, much like LSD), though i still value his righteous critique of corporate culture's desire for commercial fascism, his ability to cut through the veneer of the popular music feel-good US Festival of cultural monotony, exposing the underbelly of mafia and military tricks and treats to abolish discontent under the guise of democracy and First World enlightened civilization. at the time none of that mattered, if i could have understood the words they were beside the point. it was an eerie pleasure, sonic terror, a thrilling ride through sonic waves of simple elements combined exquisitely: guitarist East Bay Ray's goth surf boogie, bassist Klaus Flouride's amphetamine pulsations, the near-forgotten drummer Ted's manic shuffle, along with the gristling wail of Jello's would-be visionary journey through a yuppie's hellish repurposing in Pol Pot's Club Death. before the DJ gave out their name (still surprising that she would have played it at all), before i had time to wonder if this would affect the course of my life, before words came back to hold meaning, i was only aware of sound washing over me, through me, cleansing, purging, pulling me out to deeper waters and bigger waves than i had known before, a cliché now, but visceral and provocative then, sharpening my edge, nudging me towards cutting through conformity and commercial culture, pushing music where it had not gone before in my 15-year-old suburban heart and mind.

Contributor: Bob Bellerue
Age: 36
Location of memory: My bedroom, 430 El Bosque Dr, Laguna Beach, CA 92651
Year of memory: 1983/4

—

4'33", John Cage

Several years ago, Margaret Leng Tan gave a toy piano performance at P.S.122 of Cage's 4'33". It was exquisite. During the second movement, a police car from the nearby precinct on 5th street screeched around the corner, siren blaring. My reaction went from furor to amusement to wonder, as the serendipity of the moment revealed the impact of the aesthetic and metaphysical use of the glissando in modernist music, from Varese through Cage and Xenakis. I still await a more profound performance of this piece.

Contributor: Allen Weiss
Location of memory: New York

TRANSMISSION 7: EPIPHANOUS – SONIC – PASSIONS

Total imaginary monstrous otherness full of nightmares that traumatize and recast different possibilities of realization and fantasy, twisting and curving into patterns of absolute discovery–to fall over the brink and be reborn, into the living out of what could only ever be intuited, though never realized without the aid of that improvised beat caught on tape, played back in the dead of night or in the early morning, afternoons pulled out from time and stretched out on the boards of elusive categories, genre disturbed, hairstyle ruffled, and so much more.

Future Days, Can

Girl of Matches, Thee Headcoats

Bohor (1), Iannis Xenakis

The Bogus Man, Roxy Music

What Goes On, Velvet Underground

Love will destroy us in the end, Hefner

Sun Ra and his Arkestra

silence

I Put a Spell On You, Nina Simone

Third Stone from the Sun, Jimi Hendrix

A Rainbow in Curved Air, Terry Riley

She's Out of My Life, Michael Jackson

Future Days, Can

rainy night in london, homeless, wet, knock on an old hippies door. the angst of urban punk, crass, peni, flux. system problem, grrrrrrrr, guy gives a bed. lying in the corner dozing......waft of smoke, the crackle of the needle, cans future days begins. 25 mins later a musical revolution has begun. i will never think the same way again.

Contributor: Tommy Grenas
Age: 39
Location of memory: London
Year of memory: 1982

—

Girl of Matches, Thee Headcoats

I had just graduated high school and was living for the summer with my sister in a tiny flea infested apartment in New Orleans, about half an hour from where we grew up. My sister was older than me, it was her apartment but my car. One of those hot sunny summer days I was driving around New Orleans by myself listening to WTUL (Tulane's radio station) and the best song ever came on. It was bluesy. It was rocking. It was low fi and it just seemed like they meant it. I listened through something like the next 12 songs, but its college radio so they play a lot of tracks without a break. Then the news came on, followed by the jazz show. They never came on and back announced. It was the best song ever, EVER! I was stuck in traffic and this was all pre-cell phone. I was screwed. I got that song stuck in my head for years. It would pop in, show up even though I had only heard it once, I would mention this demon song to friends. The four years later I'm driving around Chicago in the cold and slush in my friend Callie's Toyota pickup listening to the tape that was seemingly always in her tape deck, one side was Bo Didley and the other thee Headcoats, a tape I had heard 40 times if once. Low and behold the song came on. IT CAME ON. At that moment space and time collapsed. Chicago was New Orleans, 1991 was 1995, the Winter was Summer, they were all one.

Contributor: Philip von Zweck
Age: 31
Location of memory: White Toyota's in New Orleans & Chicago
Year of memory: 1991 & 1995

—

Bohor (1), Iannis Xenakis

Four friends were driving, as we so often did, to the Tower Records store on the Sunset Strip in my green VW Bug. It had the most minimal of dash boards. Bare essentials, speedometer, gas gage, temperature and a monophonic AM only radio. I don't remember the classical station at the time but for some reason this time when turning it on we heard the beginnings of a complex layering of sounds emerging and reshaping eventually , after what felt like a very long time into a harsh wash of distortion. The starting sounds were crystalline gaining density in the tinny in-dash speaker, culminating in what sounded like solar winds and forest fires crackling and breaking into razor edged shards of glass. Stoned, as we were, we were transfixed and sat in the parking lot listening until it was over. When the announcer came on we took note of the title, "Electro-Acoustic Music" by Iannis Xenakis, 'Bohor (1)'.
We all found ourselves leaving Tower with that record on Nonesuch in our grimy little hands. It was an astonishment to hear something so long and unconventional on the radio. The piece remains one of my favorite electro/acoustic music works.

Contributor: Tom Recchion
Age: 51

Location of memory: Los Angeles, Hollywood, Sunset strip. In a VW Bug
Year of memory: 1971

—

The Bogus Man, Roxy Music

Waking up after a good nights sleep. A result of too many chillum pipes the night
before. This is the second day in Amsterdam. The sun is shining in through the
youth hostel window and from across the narrow street, in the hostel bar, slow
drumming and a steady bass guitar accompany a familiar voice. I somehow
know what it is but I don't know the name of the band or the title of the song.
The thickness of the music and the warmth of the sunrays engulf me and I just
lay there, free from anything that has to do with the demands of life. After a
while the need for knowing what it is gets too strong. I enter the bar and realize
that it's the new record from my favourite band.

Contributor: Carl Michael von Hausswolff
Age: 48
Location of memory: The Last Waterhole youth hostel in Amsterdam
Year of memory: 1973

—

What Goes On, The Velvet Underground

It was 1977. I was 15 years old. I had just started my first job. A teaboy (run-
ner) in a recording studio in London's Soho. I had been in bands since the age
of 11. I was pretty streetwise for a young lad but naïve when it came to drugs.
One lunchtime one of the studio engineers brought in some cannabis. I'd never
smoked it before .. and was a little apprehensive. The joint was duly rolled and
slowly passed around the room ... I took a couple of drags but couldn't really

feel anything ... in fact I didn't really know what it was I was supposed to be feeling. What I did notice though was the song that the guy who had brought in the cannabis had put on the record deck. It was the first time I had ever heard the Velvet Underground. The song was 'What Goes On'. I was entranced. I'd been into some pretty strange music around this time anyway .. stuff like Throbbing Gristle, The Residents, Cabaret Voltaire. .. but this was different .. more melodic .. yet somehow darker too. What really hit me was the wonderful, long, sinewy guitar solo. Was there one or two instruments playing the same part? It seemed to be going on forever yet I didn't want it to end ... it was almost three dimensional yet childlike in it's simplicity. At that moment it said everything to me and seemed to shine a little light as to where I could go with my own guitar playing.

Contributor: Matt Johnson
Age: 43
Location of memory: Soho, London, England
Year of memory: 1977

—

Love will destroy us in the end, Hefner

Its was summer, the summer after I finished school, and the first holiday. In Greece, driving around the volcanic island santorini, lost in its rocky landscapes, trying to find a beach. Packed in a car with 5 other friends, mid noon, the heat was unbearable; It was the type of heat that makes your eyes close and your thoughts melt. My friend played this song and suddenly a strange energy filled me, It filled all the nooks and crannies of our tight car.. 'It's the love and the truth and the hope and the faith; That will destroy us in the end' The words were popping out of the melody and all I wanted was to get destroyed, a sweet and tender destruction, by the ways described in the song. 'and I would give up everything, for a little wine, some conversation and just for being healthy'.

Feeling so confined and hot made the song so powerful, It was a moment of paranoid clarity. 'and there's no faith that will ever save me, Just from being faithful..' It was the perfect place and time and song synchronized into what i would call bliss.

Contributor: Lythia Xynogala
Age: 21
Location of memory: Santorini island, Greece
Year of memory: 2001

—

Sun Ra and his Arkestra

I turned on the radio in the middle of a music broadcast. Instantly mesmerized, I couldn't figure out who the musicians could possibly by or how they created the music. The closest I could come to answering this question was that they must be an experimental "garage band" who had taken to the practice of recording at home with each musician placed in a different room. The "distant" quality of musical coordination sounded highly unusual–the linkage between players was surely there, but their method was obscure. It was a stunning effect.

Contributor: John Bischoff
Age: 54
Location of memory: Berkeley, CA, in my home studio
Year of memory: ca. 1983

—

My special song is silence.

I have tried but I must concede that I do not have a pivotal memory or moment connected to any piece of music. Perhaps my sentiment lies in the silence in between each note or the pause between songs.

Contributor: Jessica King
Age: 33
Location of memory: Current
Year of memory: Current

—

I Put a Spell on You, Nina Simone

I am an artist who always includes music as part of her art performance/projects. I began this practice in the 1970's when there was a lot of good music around and during that time I was working on an art project: The Windmill as a Variation on Stonehenge. We installed commercial radios on the tower which held the generator and as the wind blew the generators we heard music and other information. Nina Simone's song "I Put A Spell On You" coming out of my art piece I will always remember.

Contributor: Jere Van Syoc
Age: 69
Location of memory: Grand Valley State Colleges in Allendale, MI, very near Grand Rapids, MI
Year of memory: 1975

—

Third Stone from the Sun, Jimi Hendrix Experience

In 1967, when I was in tenth grade, I went to a Halloween dance party in the garage of a ninth grader I had never met. He was playing all kinds of weird music that I was into. I didn't know anybody else in my hometown who was into such music and felt that I had entered a place of refuge from a wasteland of jocks, cheerleaders and dweebs. I went up to the host and said something like, "uh, could you play 'Third Stone from the Sun', maaan". He looked at me dumbfounded and said something like, "YOU KNOW WHO JIMI HENDRIX IS??? WHOA!!! WHO ARE YOU???" That kid was Tom Recchion. He played "Third Stone from the Sun" at my request and we have been best friends and musical collaborators ever since.

Contributor: Fredrik Nilsen
Age: 52
Location of memory: A private garage, Alhambra, California
Year of memory: 1967

—

A Rainbow in Curved Air, Terry Riley

The effects of music have always been and currently still are a great influence for me. It has taken me awhile to decide on what I should be writing about for this project. It seemed that I was dwelling on and gravitating towards a recall of an early experience, something from when I was a teenager, something that had marked me in some way, influenced me even up to today. Though, I wanted to go beyond just the song that I remembered from when I first really made out (Don MacLean's Miss America Pie), or the song from when I first slow danced (Colour My World), or the band that I first saw live (Chicago in 1969), or the band that I saw live and first smoked pot (Michael Quattro's Jam Band –Suzie's brother), or the band that I first saw and did acid (Frank Zappa or Todd Rungruen), etc., etc. Though those were wonderful moments I can vividly recall. I chose a more sober memory to share and one that perhaps has even a greater influence or transcendence for me than those more experimental moments. As a teenager in the late 60's and early 70's I listened to alternative radio in my bedroom as often as I could. I was addicted to it staying up late and calling the station to request bands like Throbbing Gristle or Mort Garson's Black Mass. Around this time I heard Terry Riley's "A Rainbow in Curved Air" on this radio station in Kansas City, perhaps KBEY were the call letters. The station would occasionally play the entire Terry Riley piece that is something around 17 minutes long. My bedroom was a perfect stage for the odd droning listening experience with colored light bulbs, batik fabrics and incense burning. I would lay motionless on my bed listening to it and be totally transcended into a timeless world of fantasy and thought. After calling the station to find out where I could find the record, I drove my mother's car further than I had ever driven it before just to find this record. My search led me to Caper's Corner, a small and alternative record store about 20 miles from the suburb where I lived. I felt like I had found an important niche for me at that moment of entering the record store and holding Terry Riley's record. Upon recalling it now, it was like a going into a portal to another world of which I have never left.

Contributor: David Schafer
Age: 49
Location of memory: Kansas City, Mo.
Year of memory: 1971

—

She's Out of My Life, Michael Jackson

My relationship to radio runs pretty deep. The gift I most remember from childhood, given to me by an older friend who doted on me when I was 7, was a radio. I listened to it constantly and particularly adored Casey Kasem and American top 40 (although I lived in Canada at the time....). And the radio is still one of my best friends – when I work on sculptures, I always like to listen to Los Angeles Lakers basketball broadcasts....in fact, I generally try to tailor my working schedule around them. Summer is a slow time in my studio. But, I digress.

As far as a singular moment, my first choice would not be a song, but the news. This was just after 9/11/2001, and nerves were running a bit high in the urban united states. I was home in LA, listening to NPR report the beginning of the Afghan war. Just at the moment they did so, the earth shook. It was a strange case of radio inciting reality. I was sure it was a bomb. It turned out to be an earthquake.

If it's songs you want: maybe, I remember crying when I first heard Michael Jackson's "She's Out of My Life". I didn't understand why I was crying, and I certainly didn't relate to the romantic feelings of the song, I was about 8. But I still love that song.

Contributor: Chris Kubick
Age: 34
Location of memory: Huntington, West Virginia
Year of memory: 1978 (?)

TRANSMISSION 8: PILLOW TALK

So much is said in a whisper...: any music can always be a love song–formed in that commingling of desire and togetherness, of the fashioning of intimate languages, defined in vocal stirrings that find vocabularies in overheard tunes, favorite melodies, the rapture and certainty found in another's lyrics. Listening and love produce their own mix-tape forever memorialized in playlists of the heart.

Save a Prayer, Duran Duran

All I want is You, U2

Bohemian Like You, The Dandy Warhols

Wonderful Tonight, Eric Clapton

Bu Bu Buana Bu (cartoon "Signor Rossi"), Franco Godi

Be My Love, Mario Lanza

Criminal, Fiona Apple

Brett

Julie with..., Brian Eno

Let's Get Together, Hayley Mills

Todo se derrumbó, Emmanuel

Save a Prayer, Duran Duran / *All I want is You*, U2 /
Bohemian Like You, The Dandy Warhols

First slow dance and first kiss: "Save a Prayer" by Duran Duran. School Dance, Gatineau, Quebec. 1983.
A University DJ once had a crush on me and told me to listen to the University radio station on the night of my birthday. At the beginning of his shift, the DJ said that the next song was for "Nadia, the coolest chick I know". The song was "All I want is You" by U2. 1993.
In a tiny bed in a tiny room in London, England. A radio was actually attached to the bed. It was the first time he and I slept together. In one of the most intense moments, "Bohemian Like You" by the Dandy Warhols was playing on the radio, and I remember him whispering in my ear, along with the song, "I like you, yeah I like you". 2004.

Contributor: Nadia Bartolini
Age: 30

—

Wonderful Tonight, Eric Clapton

It was our wedding day and things were crazy as most weddings are. Dinner was finished and now come the first dance to Wonderful Tonight. The bridal party were suppose to come in half way through the song and they were nowhere to be found. Of course I was upset and crying my husband looked me in the eyes and said. It doesn't matter if there is anyone else it's just you and me. When he said that the tears stopped and we were in this bubble of Light, Love and Warmth. He kept on saying to me it's just you and me and that was the moment that just keeps us going. So when I hear Wonderful Tonight it just brings me back to that time on the dance floor.

Contributor: Barbara Grzybowski
Age: 34
Location of memory: Wheaton IL, The Wilton Manor
Year of memory: September 9, 1995

—

Bu Bu Buana Bu (from the cartoon series "Signor Rossi"), Franco Godi

A very private joke I share with Cécile, my girlfriend, generated and refined together. One of our favourite songs of all times after that, but only works when we're together.

Contributor: Francisco López
Age: 40
Location of memory: Apartment in Montreal
Year of memory: 2002

—

Be My Love, Mario Lanza

When I was a little girl, about 8 years old, my father was a journalist in Vienna, Austria, after having not only covered the Nurenberg trials, but he also designed the American press coverage for this historic trial. My family was living in Vienna and I attended a school for American dependents. You could choose to study music with one of the Army's wives, an opera singer whose name I still remember: Mary Kay Crumbaker. While in her class, where she taught music theory, notation, sight singing to us little kids (I loved it) we had a little recital for parents. We all sang songs sitting in a circle, like Old MacDonald Had a Farm and the like, taking turns with the animal solos. Then, when we were finished, something happened that changed my life: Mary Kay appeared among us, dressed magnificently in the kimono of Madame Butterfly, and proceeded to sing "Un Bel Di" in her operatic voice, which I'd never heard. Hearing her sing was like being struck by lightening--I had no idea a human being could make such huge, vibrant beautiful sounds, and then and there I decided to become an opera singer. So, when I returned to Amercia I began looking for more music like Mary Kay's. My parents were not interested in opera, although in Vienna opera singers came to our dinner table often (they loved the plentiful food, which they, as yet, did not have access to). I found Monday night radio, to my great joy--two or three consecutive classical or semi-classical broadcasts-- one was called the Telephone Hour, and featured the great singers of that time. Dorothy Kiersten was one, and they'd have Jan Pierce, Richard Tucker, Lauritz Melchior, and many more mostly American opera singers. They'd sing from operas and operettas, mostly in English. I'd go to my room, from age 11 to about 14, each Monday night and listen alone for two or three hours to the radio, lost in a world of beautiful music, beautiful singing, and wonderful sentiments. As for remembering a specific tune, it was the one closing the program--I really don't know its title. Each singer would sing it as the final close to the program:... and now each flower/is sweeter dear. I know it's just because at last you're here. We sit alone, of all the world above/and love is brimming full/ within my heart. Later there was the Mario Lanza show which I adored. This young, handsome singer would open his program with "Be My Love", a signature song of his. I was, of course, in love with Mario Lanza, perhaps it was only his voice I loved, but that was plenty! I have a feeling you could find tapes of those old programs. They may have been on WOR, a New York station.

PS. I never became an opera singer, (I was called more strongly to the visual arts) but I have spent a lifetime studying music and voice, and I often give concerts and sing in churches. I use my voice at times in performance work.

Contributor: Fran Bull
Age: 66
Location of memory: East Orange, New Jersey
Year of memory: circa 1948-52

Criminal, Fiona Apple

For me, trying to isolate events related to music is rather hard since music has always been there in every aspect of my life. For each event there is a song, although one song will often represent several events. I would like to go off in a little bit of a tangent to make a point: The artist that made the most impact in my life is without a doubt Fred Durst. I think his music is great and he is very honest/unfiltered in what he says in his songs and in interviews. But out of everything this guy did, one thing he said in an interview really struck me and changed my way of seeing things (how much music really affected my life). He said that everybody's life is somewhat of a movie in which they are the main "actor" and the music they listen to is basically the soundtrack of their life. I think this is an awesome way of seeing things and since then, I have dated and kept every one of my personal "Soundtrack". When I have time, I sometime listen to the older Soundtracks, reminiscing at some of the ups and downs I went through at that period. Old(er) memories lost in your subconscious are often brought back, putting things in perspective. Which brings me to my point that it's surprising how music jogs your mind back in gear, makes you remember who you are, where you came from and gives you energy to take on the next phase. In the summer of 96, a buddy of mine and I were in the same boat, we both got dumped by our girlfriend and decided to go on a little road trip to Ocean City Maryland. So we packed up a tent, sleeping bags and some clothes and went cruising in my 5.0. Although we only stayed for a week, the memories from that trip are very abundant. Just the two of us on a rampage in a strange town, the parties with total strangers, the sun, the beach, seadooing and the nightlife. Up to today, that was probably the best time I've ever had. Throughout the week, the one song that kept playing on the radio was "Criminal" by Fiona Apple. Even though it's been 8 years, that song always jets my mind right back to MD and that summer of 96. The memories do not age or go out of style. Hopefully Ms. Apple, Mr. Durst and all artists realize how much they touch our every day life and how important they are to us.

Contributor: Francois Cuillerier
Age: 32
Location of memory: Ocean City, Maryland
Year of memory: 1996

—

Brett

Two years ago, the love of my life made me a mix tape on which he included a melodic tune written and sung by a dear friend singer-songwriter of his. After this supposed love hurt me drastically, I continued to listen to this mix tape and was particularly intrigued by the slow, moving tunes of his guitar-playing friend. The song that his friend sang was a song wrought in bittersweet heartache and love pain. Two years later, I was standing amidst a crowd of over 5000 people in a musical festival called Power to the Peaceful in San Francisco's Golden Gate Park. There were tons of small bands playing everywhere with a headliner on the huge stage barely visible to the majority of the attending public. Amidst the cacophony of different sounds and beats, I heard this singer-songwriter-friend-of-exboyfriend singing his heartbreak song sweeter than a bird in June. I let my ear follow the tunes until I was standing a mere 3-feet away from the young music man. All of a sudden, I started to cry. After I had spent days and months and now years getting over my hurt, it all came flooding back when the music knocked me right back to the suffering place I had been standing on back in the days of the mix tapes.

Contributor: Jennifer Colker
Age: 22
Location of memory: Golden Gate Park, San Francisco, CA
Year of memory: 2004

—

Julie with..., Brian Eno

There are many incidents associated with songs and music...music and sound is
the fabric life. But many of the stories that leap to mind of memorable experiences with music are about the music itself...the revelatory excitement it gave
me when I experienced, for the first time, the music I always wanted or needed to hear: The Residents, "Anarchy in the UK", Ethiopiques, Einstuerzende
Neubauten... Or transcendant, exultant experiences where I heard music that
transported me, made me want to ascend thru the ceiling, made the hairs on
my neck stand on end : Steve Reich's "Drumming" performed live, Kraftwerk's
"Radioactivity", This Heat or Glenn Branca or Muse or The Birthday Party per-
forming live, freq_out in Norway! So many more... But one incident I remember
with direct reflection on events in my life, which caused a tugging in my hear,
a lump in my throat and a synchronicity, was the song "...Julie with" by Brian
Eno from Before and After Science, when I had just broken up with the first love
of my life named "...Julie". Ah, mon agonie douce.

Contributor: J.G.Thirlwell
Age: Currently unavailable
Location of memory Melbourne, Australia
Year of memory: 1978

—

Let's Get Together, Hayley Mills

Hearing Hayley Mills' pop hit "Let's Get Together" which was released in con-
junction with her hit Disney movie "Parent Trap." I was 11 years old at the time
and I had a total crush on Hayley Mills. I remember being intensely love-sick
hearing the song.

Contributor: John Bischoff
Age: 54
Location of memory: Piedmont, CA, in my room in my childhood home
Year of memory: ca. 1961

—

Todo se derrumbó, Emmanuel

When I was 17 years old I had my first girl friend (she was 15), after almost 1 year of relationship, in December 1980, when I was already 18, I was on a Christmas vacation trip with a friend and my brother in the south east part of Mexico, through the Mayan ruins and the beautiful Caribbean beaches. There was a song playing all the time from an LP, a famous Mexican romantic POP singer called Emmanuel, the song went "Todo se derrumbo dentro de mi, dentro de mi. De humo fue tu amor, y de papel, y de papel" (Everything felled apart inside me, made out of smoke was your love, and out of paper). After I came form the trip, I saw her and she ended up the relationship. I got the record form a store and put the song all the time, suffering from this terrible outbrake .

Contributor: Manuel Rocha Iturbide
Age: 41
Location of memory: Various cities in the south east of Mexico, and then in Mexico City
Year of memory: 1980-81

TRANSMISSION 9: BLIND ALLEYS & DEAD ENDS

Urban journeys self-styled by memories of overheard histories, stories of imagined possibilities that your steps aim to draw into being, knowing the futile end, the buried dream beyond recovery. Realized in fights: of fists or emotion or languages scrawled out in diaries of hope, regret, and obsession. And retold in audible trajectories across musical time-zones.

Common People, Pulp

Great Balls of Fire, Jerry Lee Lewis

Coucouroucou Paloma, Caetano Veloso

Papa Was a Rolling Stone, The Temptations

Show Me The Way, Peter Frampton

Unfinished Sympathy, Massive Attack

Summer Breeze, Seals & Crofts

Common People, Pulp

While approaching the Shibuya crossing in Tokyo, they were playing the video of Pulp's "Common People." As I got to the crossing and watch their video and then I looked across the sidewalk, I was facing at least a million common folks going to the other side of the road. It was a strange moment for me, and somehow and to this day, that song is incredibly moving to me. Also it was odd to hear a song with the visuals in the middle of a busy intersection.

Contributor: Tosh Berman
Age: 50
Location of memory: Shibuya, Tokyo Japan
Year of memory: 1995

—

Great Balls of Fire, Jerry Lee Lewis

I got this request from Brandon LaBelle to recount a memory of a significant moment in my life connected to a particular song. I thought to myself, "why send Brandon some ratty old memory, occluded by years of dust and full of mnemonic moth holes?" I decided to make a new memory, custom-designed for Brandon's project. "Even better," I thought to myself, "instead of storing the memory in my dilapidated brain which often returns 404-File Not Found errors, I could store my memory in Brandon's library/database where I know it will be safe until I need it again." So, this afternoon, I took the 345 bus from near my house in Clapham to the King's Road in Chelsea. If you know your punk rock history, you know that the King's Road is where the original London punks used to hang out. In fact it was at Malcolm McLaren's and Vivienne Westwood's shop, "Sex", where John Lydon first auditioned for the Sex Pistols by singing along to the Modern Lovers' "Roadrunner" on the jukebox. I walked the length of the King's Road, trying to feel the ghosts of those halcyon days, when punk was not yet a brand name, co-opted by high street department stores. But, as has always been the case when I've walked the King's Road, I feel no ghosts and am confronted instead by the fact that the King's Road is just a road and that punk, as something worth caring about, has come and gone. No use crying over spilt hair dye. I passed over a couple of traditional English "cafs", as they're known, and turned, like a man turning into a diner, into Ed's American-style diner. Having grown up in America, much of that time in diners, I thought to myself "where better to make a memory for Brandon's project than here, in an American diner on the King's Road, birthplace of punk?" The place is transparently kitsch, all 50s roadside diner linoleum tiles and vinyl booth benches. At each table a genuine 1950s jukebox provides the piece de resistance of the bald-faced lie of the place. These jukeboxes that once took American coins and played 7-inch, 45 R.P.M. vinyl records, had been retrofitted to take English 20p pieces and to send a signal to a CD player deep in the bowels of Ed's American Diner. On each juke box, a selection of American 50s and 60s rock and roll, ranging from Elvis Presley's "Hound Dog" to the Beach Boys' "Surfin' USA". I ordered the classic beef burger medium-rare, only to be informed that they do them medium, medium-well, or well-done. Health and safety. So I settled for medium and fries and a chocolate malt. I like chocolate malts... I examined the jukebox choices carefully, wanting to finish off my memorable visit to the theme-park American 50s diner with the perfect American 50s soundtrack. I rejected "Hound Dog". The song has become so emblematic, so iconic, that, in a way, I think we've all become deaf to it. I rejected "Surfin' USA" too. I was after the 50s, not the 60s. And, besides, I never much liked the Beach Boys. To complete the memory, I popped my 20p piece into the juke box and I punched the F button and the 7 button. F7 is "Great Balls of Fire" by Jerry Lee Lewis. Among all the original rock and rollers who recorded for Sun Records in Memphis in the 50s – Elvis, Carl Perkins, Johnny Cash – Jerry Lee Lewis was the closest in spirit to the London punks of the 70s. He was a musical anarchist. He played the piano with his feet. He had an incredible fountain of curly hair that he combed and coerced into a slicked-back pompadour. But when Jerry Lee let loose at the keyboard – as he was wont to do – that hair erupted from his head like a flame leaping from pile of junkyard tires. I ate my burger and sipped my thick malt through my striped bendy-straw. Wild, incongruous hoots and hollers emanated from muddy speakers buried in a wall or under the counter at Ed's American Diner. I thought to myself, "someday I will look back on this and smile."

Contributor: Seth Kim-Cohen
Age: 40
Location of memory: Ed's Diner, King's Road, Chelsea, London
Year of memory: 2004

—

Coucouroucou Paloma, Caetano Veloso's version

I was sitting in a cafe talking with my lover when Caetano Veloso's version of Coucouroucou Paloma was played over the sound system. We were speaking in generalities about relationships. This song is a beautiful Mexican song of longing. It seemed to me as if it was speaking to us about our longing and the fragility of our affair, of its impossibility even though at its inception it seemed that we had crossed so many barriers to begin with. I looked over at her watching her breathe as she looked down at her hands. What is one to think of the breath that escapes from their lover's mouth, of the intimacies shared? It all seemed so impossible as if each moment we were under erasure, so frail. It made me think of Almodovar's film Talk To Her. This song also appears in this movie. A movie about impossible loves, impossible attractions and as in other Almodovar films the production of subjectivity in the sense of Deleuze and Guattari and in the case of Almodovar how it constructs libidinal flows in relation to sexual preference and social mores. There I was holding the hand of my lover a lesbian as if we were in an Almodovar film being serenaded by Veloso singing us this song of longing for our impossible love.

Contributor: Michael Weinberg

100

Age: 41
Location of memory: Los Angeles in a cafe
Year of memory: October 2004

—

Papa Was a Rolling Stone, The Temptations

Sitting in the back of a baby blue Ford Falcon Station Wagon on a snowy Upstate New York night mid Dec 1972 waiting for my father to collect some chump
change on an outstanding balance for an Electrolux vacuum cleaner he had sold
to a single mother who didn't need it and couldn't afford it.

Contributor: Lydia Lunch
Age: 45
Location of memory: Rochester NY
Year of memory: 1972

—

Show Me The Way, Peter Frampton

I had just moved back to Los Angeles from living in Denver for one year. I was
in the 10th grade at North Hollywood high school. I don't even remember the
class this memory is attached to, but I do remember the song and the person
who kept telling me how "You gotta hear this song, it's so fucking rad, it will
change your life." This person's name was Cindy. She used to sit in class breaking apart blue-colored diet capsules. She explained to me, "The pink balls are
speed and the white balls are sugar. I don't want the white balls 'cause they're
bad for your teeth, so I pick out the pink balls." She spent most of the class
hour separating the pink from the white balls. She could separate five capsules'
worth of pink balls in the course of one class period. Each capsule yielded be-

tween 20 to 25 pink balls. Cindy sat towards the back of the class and kept her purse up on the desk, so the teacher didn't notice her. I sat next to Cindy. Cindy had peroxide blonde hair, parted in the middle and long down to her waist. She usually wore tight Levis bellbottom cords, a halter top and Wallabies. Needless to say of someone with an amphetamine habit, she was bone thin. Her skin was so white it seemed to glow. She wore tons of make-up, nearly in done-up chola style but still respectfully detached from this: she was white and lived in an apartment house behind the Thrifties on the corner of Laurel Canyon and Ventura Boulevards. She smelled of menthol cigarettes. I wasn't in love with her but she fascinated me. When Cindy had enough of the pink balls she would crush them to a powder on the formica desktop with the bottom of a nail polish bottle. "Now, this is the best part," she said, as she swept the pulverized pink diet balls onto her make-up mirror. She took a dollar bill out of her purse, rolled it tight and inhaled the pink powder in one pass. Her timing was so perfect that the class bell rang five minutes after she'd finished. "You mean you don't know Peter Frampton? Get serious! 'Show Me the Way,' dude! I mean, like, he's playing guitar and, you know, like, he's singing and what he's playing on the guitar is what he's singing, so it's like he's singing on the guitar too. Way rad! And a total fox! What radio station do you listen to, anyways?"

Contributor: Jason Kahn
Age: 44
Location of memory: Los Angeles, California, North Hollywood High School, Classroom
Year of memory: 1975

—

Unfinished Sympathy, Massive Attack

My brother* had just come back from Holland. Quite a metallic start to this one, but very unmistakeable. I can ....we had been drinking earlier in Nancy

Spain's bar on Barrack Street. I* remember a drink spilling on our table [a tree trunk cut through] and unknown to me the beer had drained from the table onto my seat, soaking me with alcohol. I was so warm and cloudy inside that I was unsure whether I had urinated on myself or not. [I asked my friend Eoin*, and he told me that I probably had]. . . Later in Henry's* [One of the first times in a nightclub dancing to techno music.] I had an intense sense of the oneness of the world, how the blood in our veins is all essentially the same blood. People smiling and laughing, everybody breathing the perfume of each other's sweat – holding each other…laughing and crying – sometimes simultaneously. Something of the entwining of the orchestral and the vocal in this number so numbed me to anything that was outside of this particular [I hugged my brother and told him I loved him.] moment. Soaring seamless. Elevated, upward, forever… Cut me loose! Jettison the weights for Christ's sake! Let me go! Let me go!

I was so overcome I had to sit down. Little planks of ply thrown together near the window overlooking North Main Street. Directly below I could see about fifteen local youths repeatedly kicking and punching a single person to within an inch of his life. Like a swarm… the geometry was compelling. Waves of bodies in and out in unison, [queuing for a turn to have a go]. A mini tornado with a limp quietness at it's centre. It was the most ferocious group act of violence that I have ever witnessed. Nobody dared help him, for fear of getting caught up in the maelstrom. They had seen it all before… some pitiful local drug feud I was told.
[He told me later he always thought it was called unfinished symphony.]

*My brother Patrick – recently divorced [three children] and living in Boston, USA. I have found it difficult to talk to him recently after he broke my mothers left index finger in an alcoholic rage.
*Me, James O'Leary, currently sitting in front of my computer in a flat in Hackney, London. Just this morning finished reading a book called memoirs.- a book about memory and longing.
*Eoin, an old friend of mine – had his first baby last Friday [named Juno]. He

lives in Ann Arbor, Michigan, USA. We have not spoken in some time.
*Sir Henrys nightclub – recently demolished, currently a building site for riverside apartments.

Contributor: James O'leary
Age: 31
Location of memory: Sir Henrys nightclub and environs, North Main St., Cork City, Ireland
Year of memory:1991?

—

Summer Breeze, Seals & Crofts

The accident was as vivid in my mind as if it had happened last week. But the date was hazy until I opened a Billboard chart book and saw that "Summer Breeze" by Seals & Crofts had entered the Hot 100 in September 1972. So, yes, it must have been late Fall '72. I was 21 years old and living in the university section of Minneapolis. But one cold, clear Saturday, I found myself miles from town, an unknown party guest on a stranger's farm. I remember the humid atmosphere in the rundown farmhouse, and the view of the bare, frozen fields from the kitchen window. There was reefer, there was whiskey, there was a keg of beer and some pills of the "downer" strain. The day passed, and early dark was falling. Somebody mentioned a ride back to town, and I climbed into a crowded Volkswagen van. I recognized the driver as someone I'd seen "knocking back Jack" (Daniels) throughout the afternoon. We started down the dirt road towards the highway, travelling at perhaps eight per hour. Less than a mile from the house, the van gently eased off the road and fell on its side in a shallow ditch. Inside, passenger reaction was not one of panic but of surprise ("Hey!"). This was followed by disappointment ("Shit!") at the realization that now we'd be forced to awkwardly climb out of the van, then walk back to the house in the cold and dark. We did, hunched and shivering against the wind off the fields, and eventually I found another ride back to Minneapolis in another vehicle. As the car radio played, the DJ's metallic voice scraped against my adrenaline, my regret and relief. "Summer Breeze" is the song that played as I checked again for bumps and bruises, feeling lucky to be alive.

Contributor: Andy Schwartz
Age: 53
Location of memory: Rural suburb of Minneapolis, Minnesota
Year of memory: 1972

TRANSMISSION 10: ETCETERA

To provide a place for the left-over, the unlocatable, the driftwood of the radio-ocean of song and memory, for there must be a space built for the unclassifiable, for the periphery of any system. Here, transmission creates its own outline, its own narrative, concocted from various stabs in the dark, tangents of noise and other spurious routes.

Morena de Angola, Chico Buarque

Abracadabra, Steve Miller Band

Keith Jarrett

El Gusanito, Jorge de la Vega

From Her to Eternity, Nick Cave & The Bad Seeds

pause signals, Norwegian radio

Suffragette City, David Bowie

talk show

Goldberg Variations, Bach

Morena de Angola, Chico Buarque, sang by Clara Nunes

I was four years old, went in the middle of the night to my parents bed, probably scared with something, slept the rest of the night there and woke up with this song on the radio, my mom and dad (also one of the few memories I have from the time they were together) getting ready for work, bath set, open closed, both of them coming back and forth, sometimes they sat on the head of the bed, and Clara Nunes, with flowers on her hair and a white dress inside the radio. I still couldn't understand how could she be inside the radio! This lasted much more than it's 3 or 4 minutes, it's been lasting for over 25 years.

Contributor: Fernanda Farah
Age: 29
Location of memory: Curitiba, Brazil, at my parent's room
Year of memory: 1980

—

Abracadabra, Steve Miller Band

First song my sister and I taped from the radio to our tape recorder: "Abracadabra" by the Steve Miller Band.

Contributor: Nadia Bartolini
Age: 30
Location of memory: My sister's bedroom, Gatineau, Quebec
Year of memory: 1982

—

Keith Jarrett

dear brandon – hope the project goes well – here is my contribution (there could
have been many, because I listen to the radio through the night, so quite a lot
of music and voices are associated with surfacing from sleep, darkness, drifting and moving into morning): I woke at three in the morning to Keith Jarrett
playing the Köln concert and then played it on CD watching the sun disappear,
a comet in the sky.

Contributor: Jane Calow

—

El Gusanito, Jorge de la Vega

This song synthesizes a story that has developed in many years and steps. When
I was at school they brought me to an exposition of an Argentinean group of
artists called "Nueva Figuración". I didn't know much of this movement, but
there was a special thing that kept in my mind. It was something like a poem:
"El Gusanito", by Jorge de la Vega. An absurd idea that captured my attention.
Years later, someone told me the poem was part of a song. There was another
piece in the puzzle of this story that was only a resonance by that time. The issue
appeared several times along that period. One day, in the morning, I was hearing a radio program at home in "Radio de la Ciudad". Suddenly, the presenter
sang "the" song. Yes, it was the poem! And it was very beautiful! It was new for
me (but not too much), and I wanted so much listening to it! Later I listened to
the song performed by his author, de la Vega, in the radio too. The line of the
story didn't conclude, it opened to other sides. I like very much this ironic and
melancholic song.

Contributor: Camila Juárez
Age: 30

Location of memory: Buenos Aires, Argentina
Year of memory: 1989-2003

—

From Her to Eternity, Nick Cave and the Bad Seeds

The 29th of April 1994, a Friday night, Nick Cave and the Bad Seeds gave me
a life long Tinnitus. They were giving a concert at the Palladium located at
Kungsgatan (the Kings Street) in Stockholm. I was there in the audience. The
solo guitar and the cymbals were so high pitched through the PA that it gave
me an ear hiss that never stopped. I have to live with that sound no one else can
hear for the rest of my life. I suppose it is a gift.

Contributor: Leif Elggren
Location of memory: Stockholm
Year of memory: 1994

—

pause signals, Norwegian radio

You see my radio memory which is strongest is from my child hood: the pause
signal, between broadcastings, when I was ill, lying in bed, not in kindergar-
ten, and also the stock exchange and weather reports from the fish banks in
the north sea ... strangely enough I think a lot of people in my generation have
this same memory of these sounds...I once used the pause signal in a piece. No
particular song...

Contributor: Jana Winderen

—

Suffragette City, David Bowie

the shock of david bowies androgenous alien figure striking like a techno mantis from the tv screen, and creating the first feelings of awe and the other world of music lurking behind the door. making me feel very drab in my school uniform, garnished with irn-bru and corned beef sandwich.

Contributor: Tommy Grenas
Age: 40

—

Talk Show

I never listened to music on the radio when I was a kid growing up in the Pacific Northwest. From the age of 8 till 18 I only listened to talk radio. I remember, when I was 12, listening to a young woman describe in detail how she was raped while hitchhiking. I also remember, around the same time, hearing this one guy discuss how he was from another planet (Saturn to be exact). And not only was he from another world, but that he was still on that planet. He was merely projecting an image of himself into the radio studio. I didn't believe a word of it, but I did love the poetry of his lexicon.

Contributor: GX Jupitter-Larsen

—

Goldberg Variations, Bach

Just a little anecdote from life in London: I exited London Bridge station on my way to work. As usual I passed a curiously small (an archway the size of a caravan) English cafe that is styled like something from the 40s, but in a very subtle way. I went in this time and had toasted bacon and egg sandwich with a cuppa tea! Something I never have. How very English, yet how good this tasted. The bread seemed Italian though (as did the owner) and I have to admit a little more care was taken than your average caf!! The owner had the radio on and suddenly I was set into a dreamy world as though looking through strangely shaped glass panes that blurred and focused all at the same time. It was Bach's Goldberg Variations (Aria?) played so slowly and tenderly. Do such memories fade?

Contributor: Mark Schreiber
Location of memory: London
Year of memory: 2006

Archipel Festival, Geneva (March 2005)

Ten mobile sculptures with interior playback systems. Each sculpture represents a section of the radio memory library, and is located in the festival space with the intention of allowing visitors to move them around and use them as they wish. A CD of the songs related to the memories are played on loop-mode, and appear as background music/noise to the functioning of the festival.

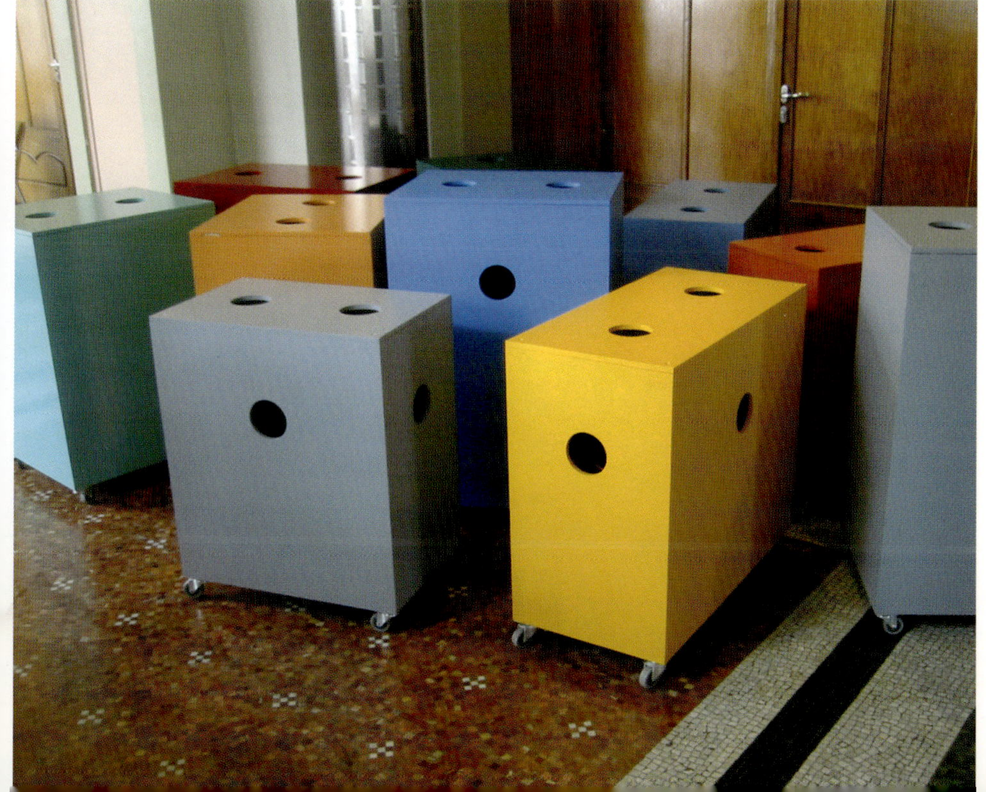

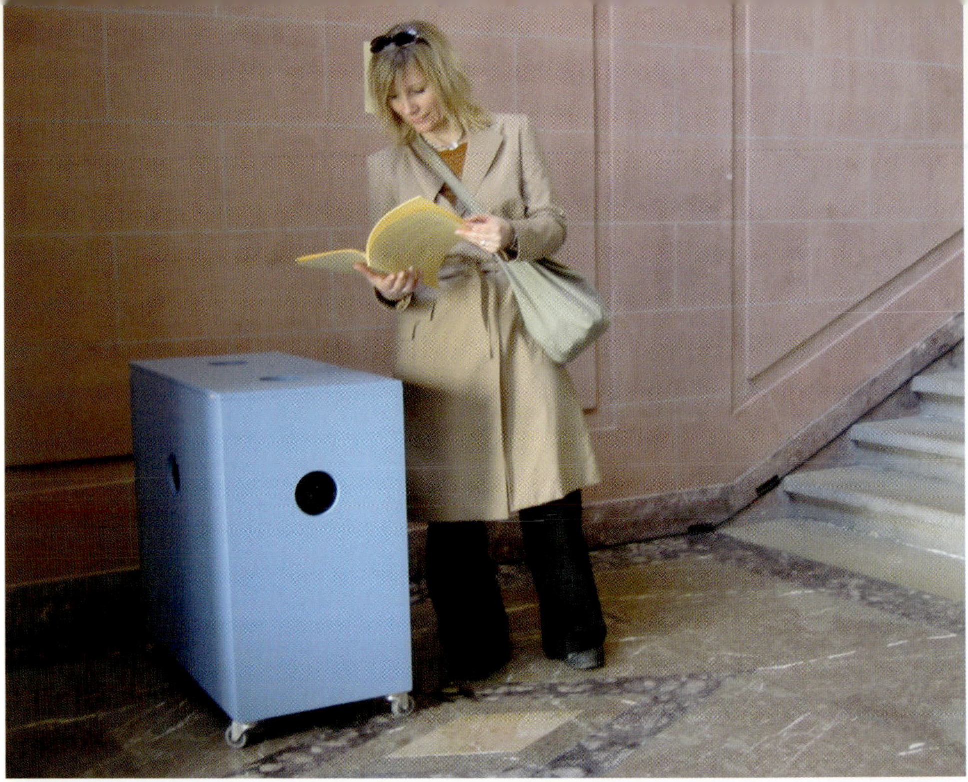

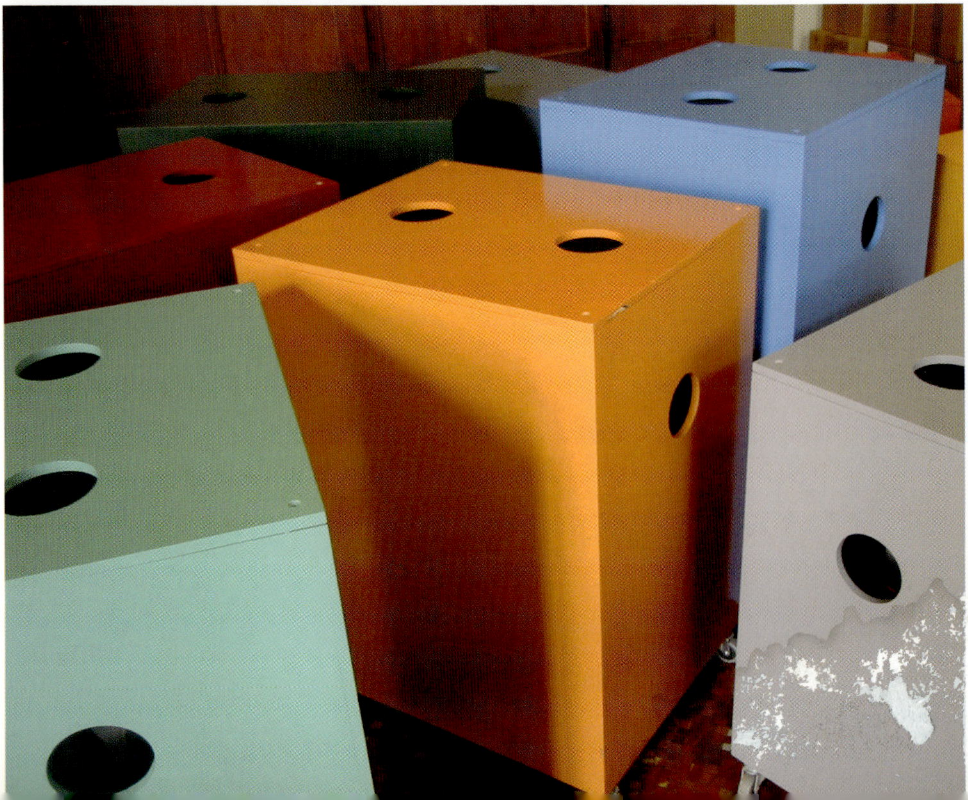

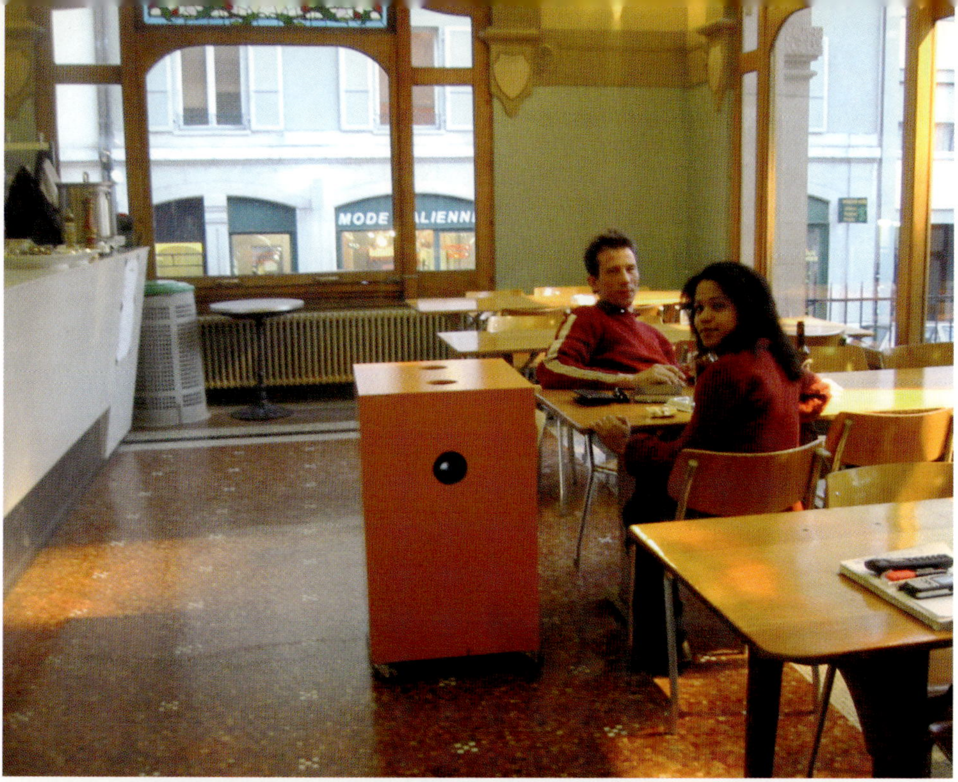

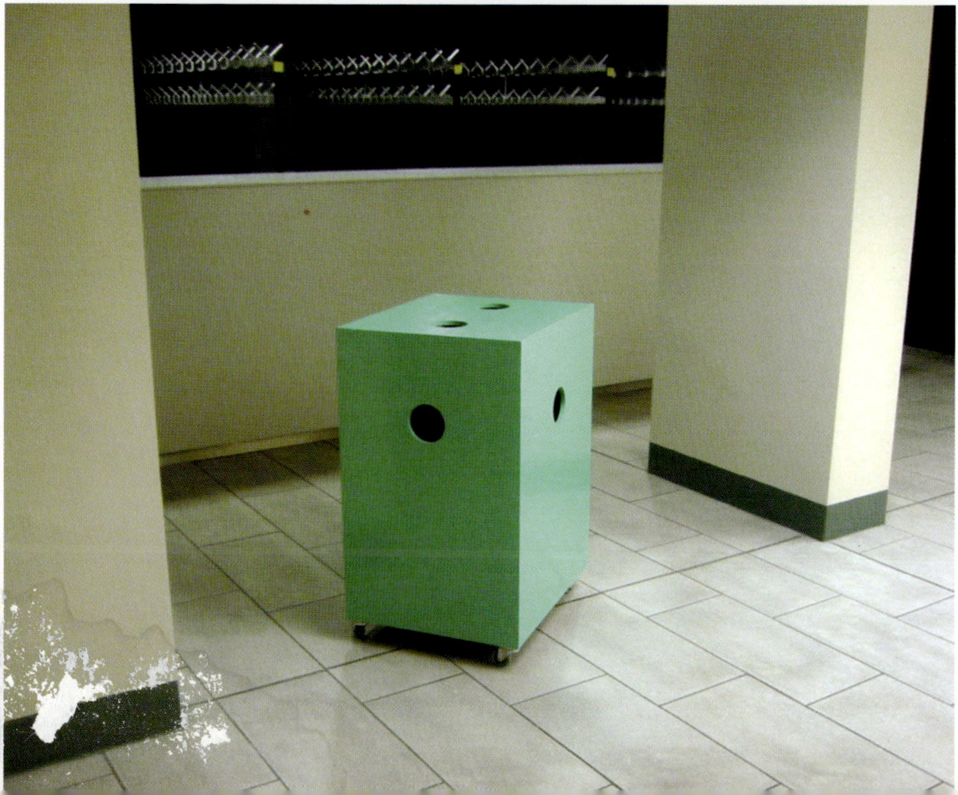

Espace Gantner Gallery, Bourogne (February 2005)

Ten tables with portable CD players and battery-powered speakers amplifying the radio memory library, accompanied with booklets of the memories. Visitors experience the audible overlap of the looping CDs, as a room noise, while reading through quietly to themselves.

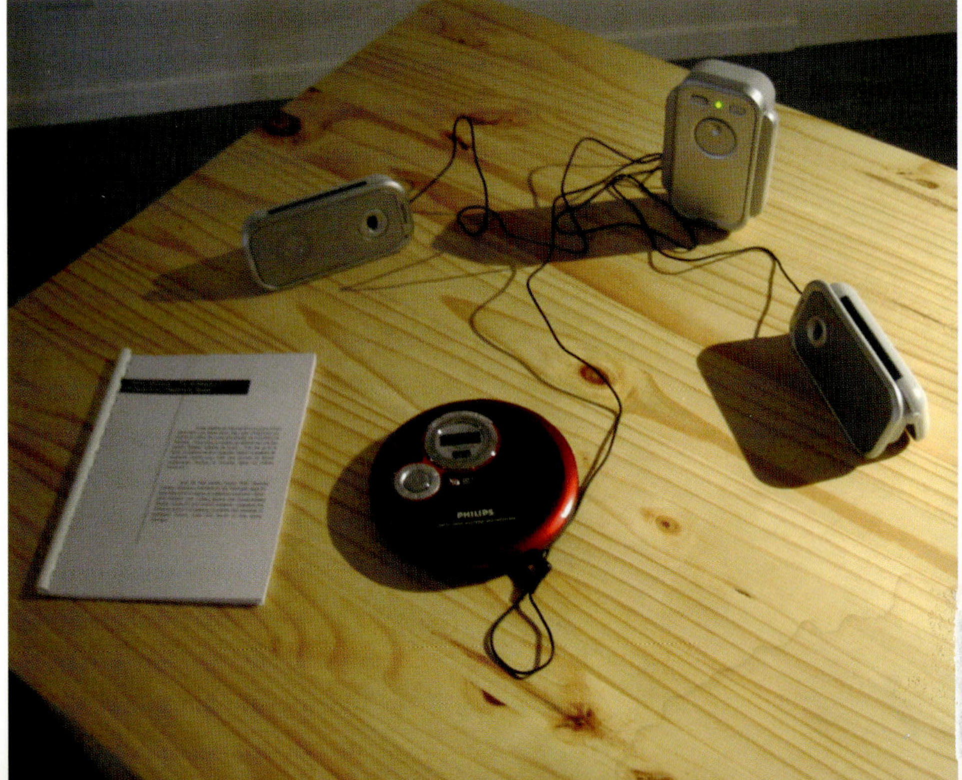

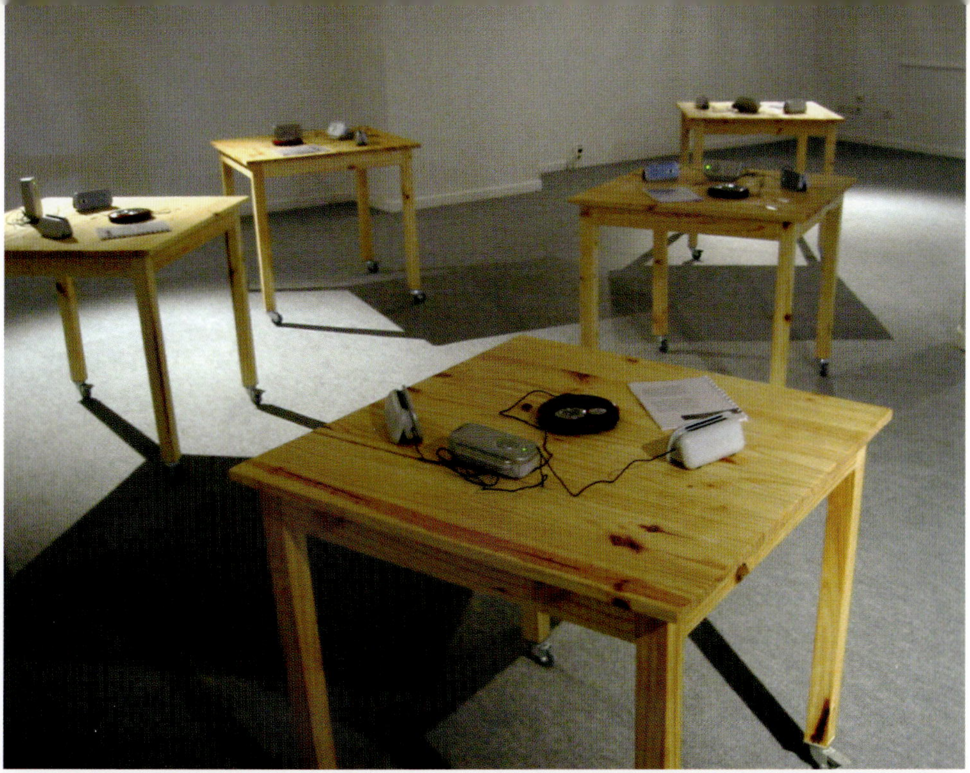

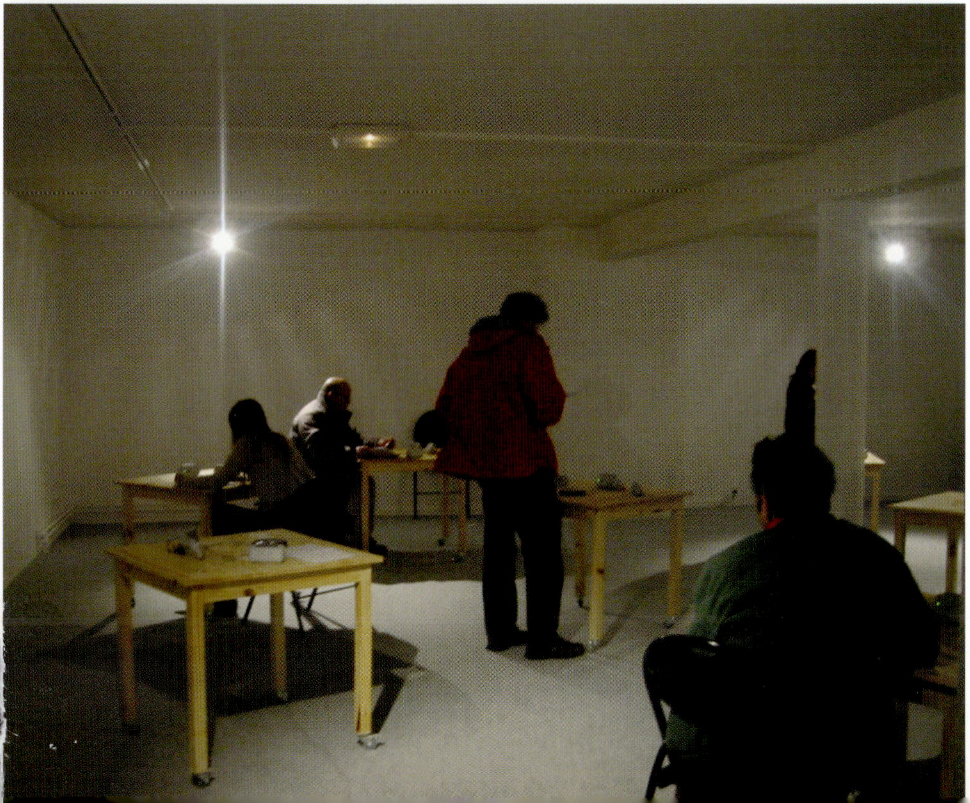

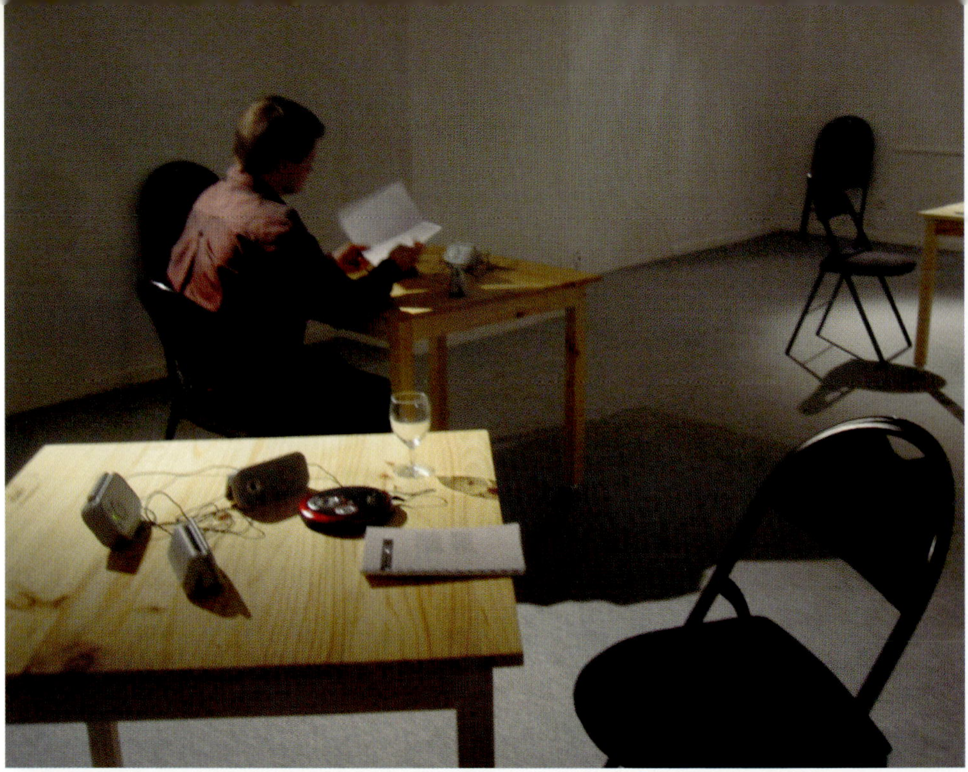

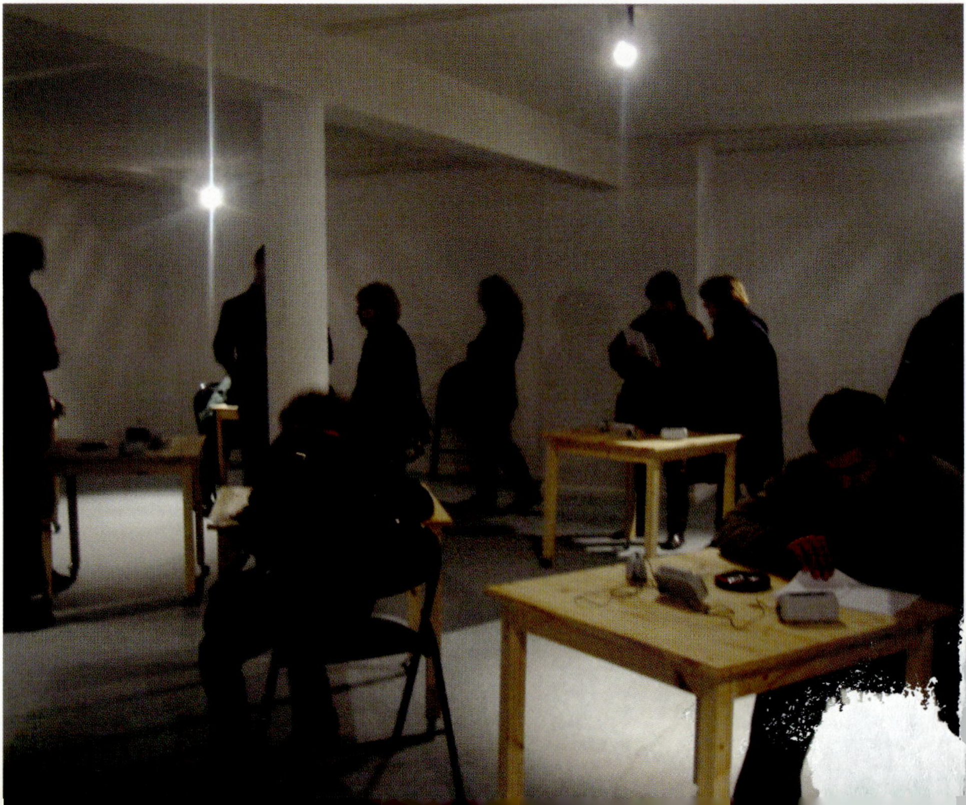

Radio Revolten, Halle (September 2006)

Locating the radio memory library in a local cafe in the city of Halle in Germany, which functions as a social space for the Radio Revolten festival. The cafe acts as a hub, and the playing of CDs through seven players and speaker systems turns the cafe into a public jukebox – visitors can turn the players on and off as they wish, DJing the space throughout the day. Booklets with the radio memory library are placed on the tables for people to read through. In addition, tablecloths are placed on the tables, and visitors are invited to add their own radio memories by writing directly onto the cloths. These are collected and added to the library.

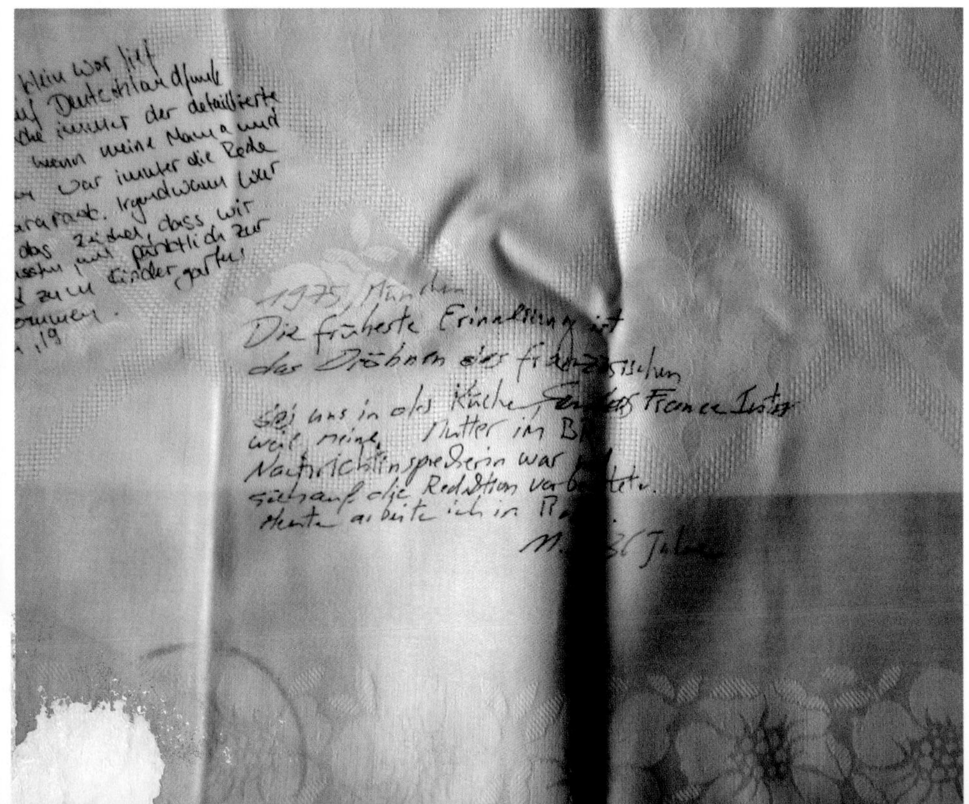

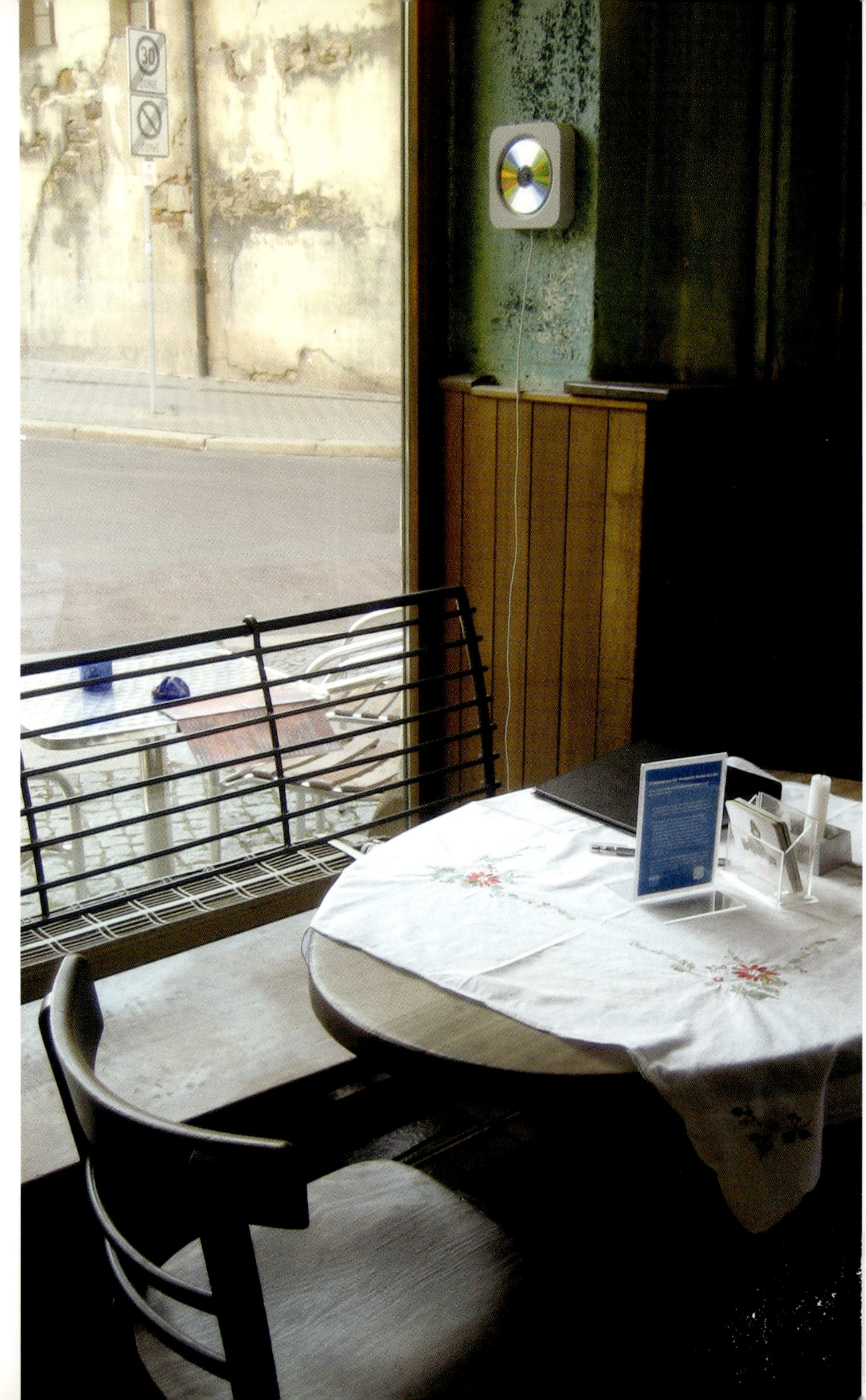

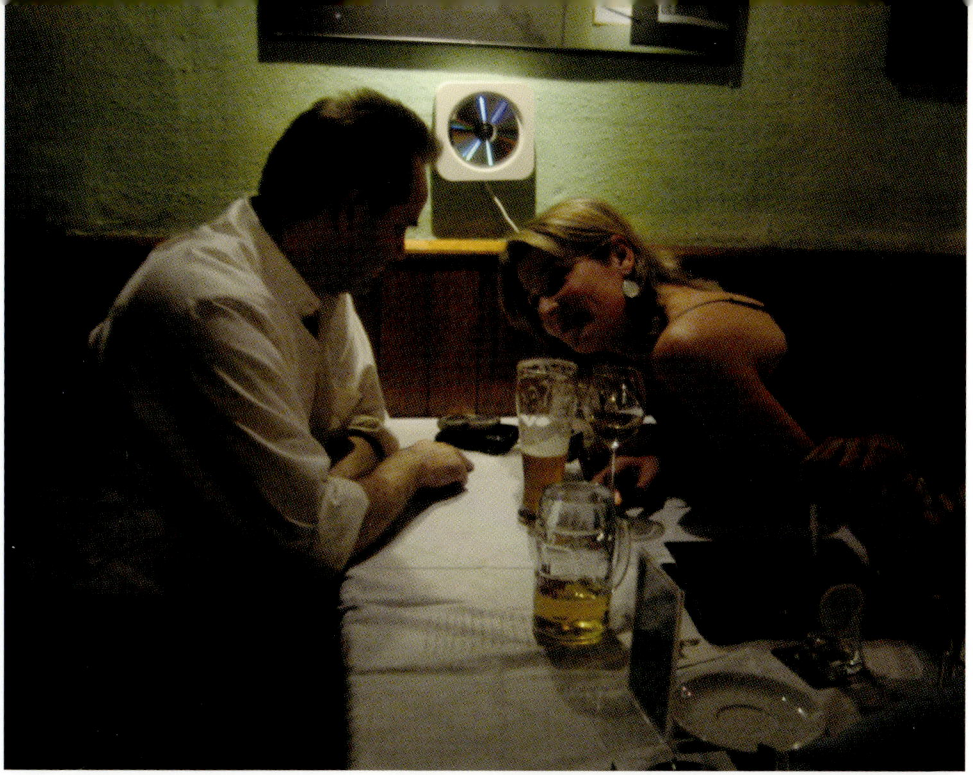

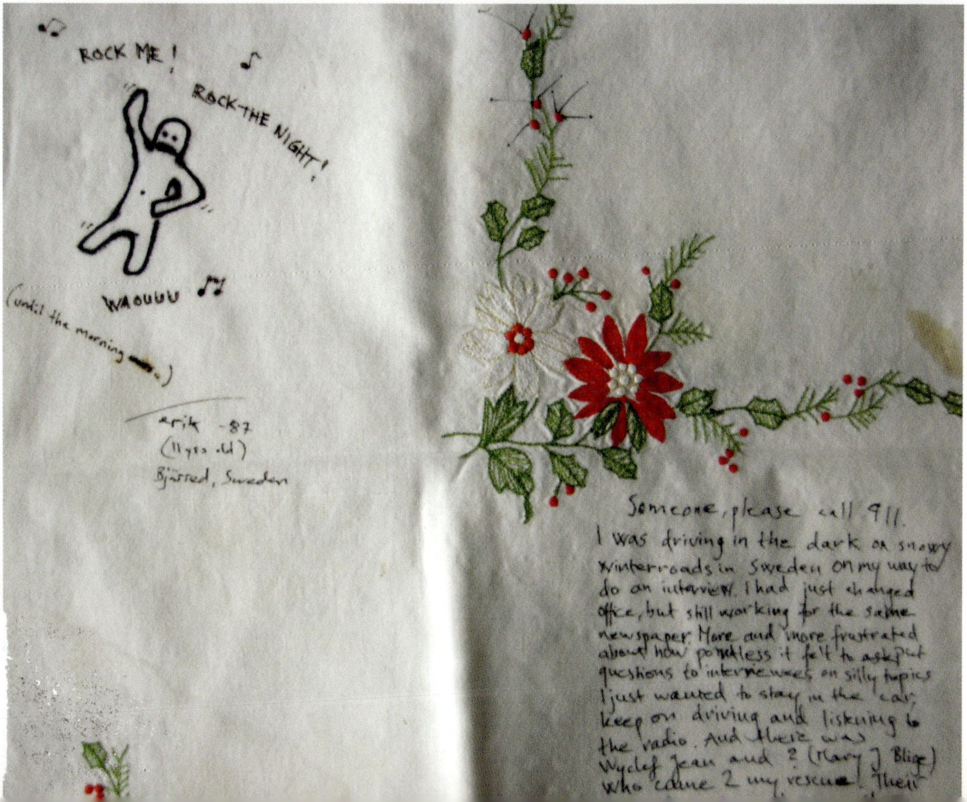

ROCK ME !
ROCK-THE NIGHT !
WAOUUU
(until the morning —)
erik -87
(11 yrs old)
Bjärred, Sweden
Someone, please call 911.
I was driving in the dark on snowy
winterroads in Sweden on my way to
do an interview. I had just changed
office, but still working for the same
newspaper. More and more frustrated
about how pointless it felt to ask put
questions to interviewees on silly topics
I just wanted to stay in the car,
keep on driving and listening to
the radio. And there was
Wyclef Jean and ? (Mary J Blige)
who came 2 my rescue. Their

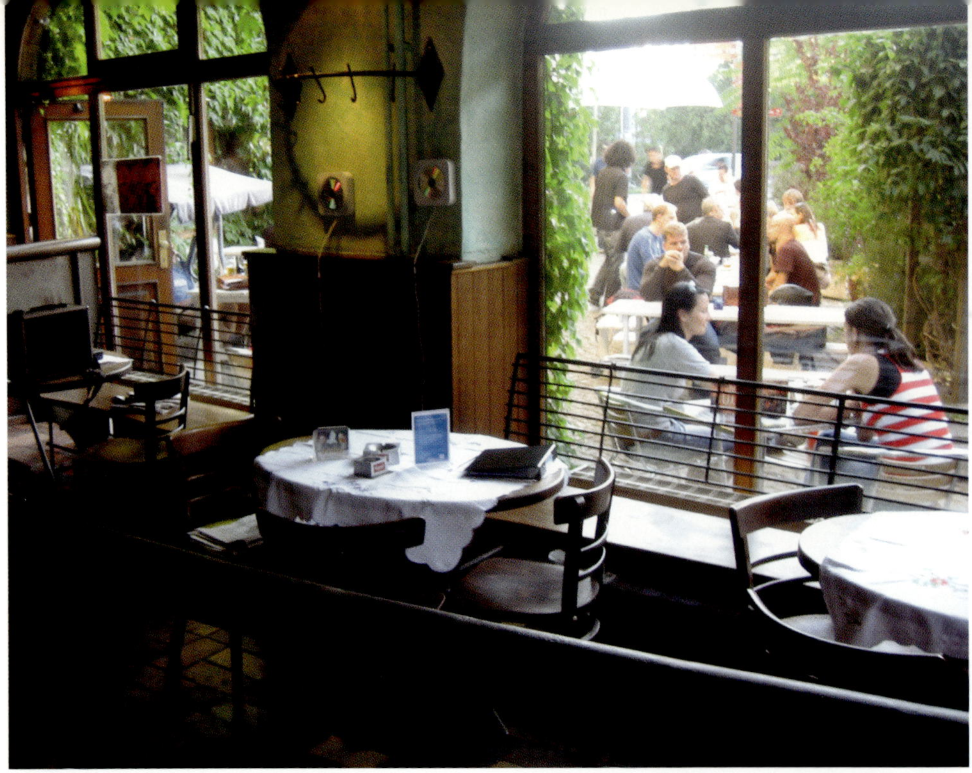

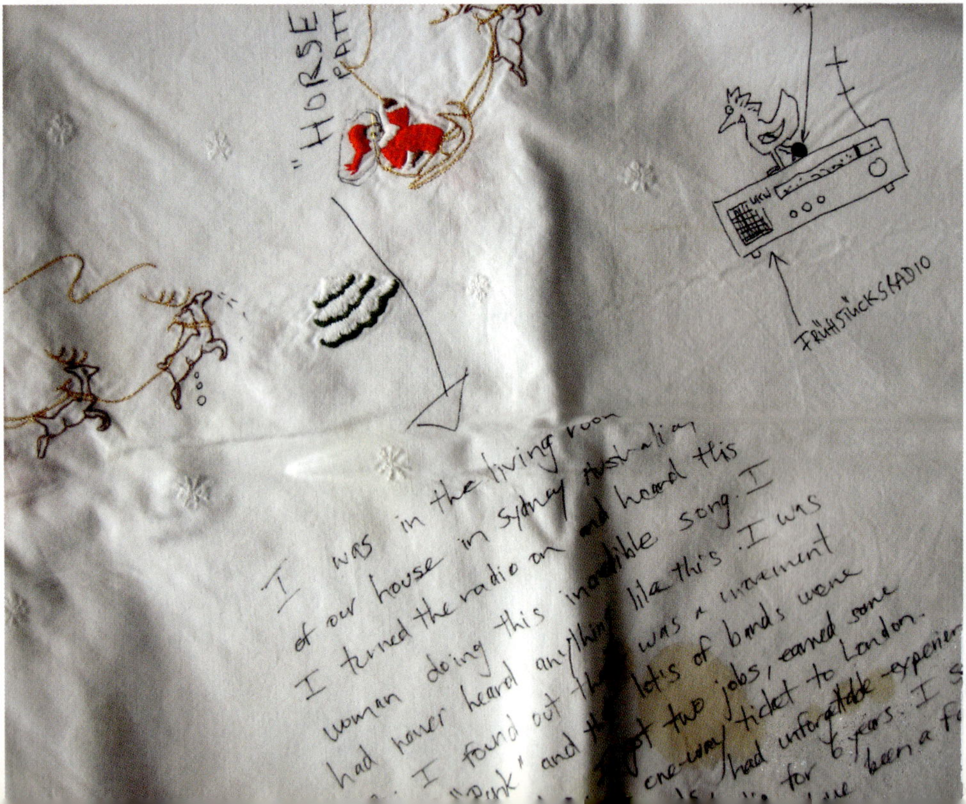

"HORSE
FRÜHSTÜCKSRADIO
UKW

I was in the living room
of our house in Sydney Australia
I turned the radio on and heard this
woman doing this incredible song. I
had never heard anything like this. I was
... I found out the ... was a movement
... "Pink" and the ... lots of bands were
... got two jobs, earned some
one-way ticket to London.
... had unforgettable experiences
for 6 years. I s...
... been a f...

Lydgalleriet, Bergen (April 2008)

Working with four memories from the radio memory library, a series of architectural models are made, which try to build up the scene of the memory, and turned into videos with accompanying soundtracks giving a spatial and atmospheric expression to the memories. These are presented as part an installation that stages, through dramatical setting, the four memories, with tables, chairs, lighting. Expanding on this idea, the memories are also turned into written scripts – dialogues between friends, secret monologues, an imagined conversation with a radio... – creating extended narratives from the original story, which are presented to visitors for enactment.

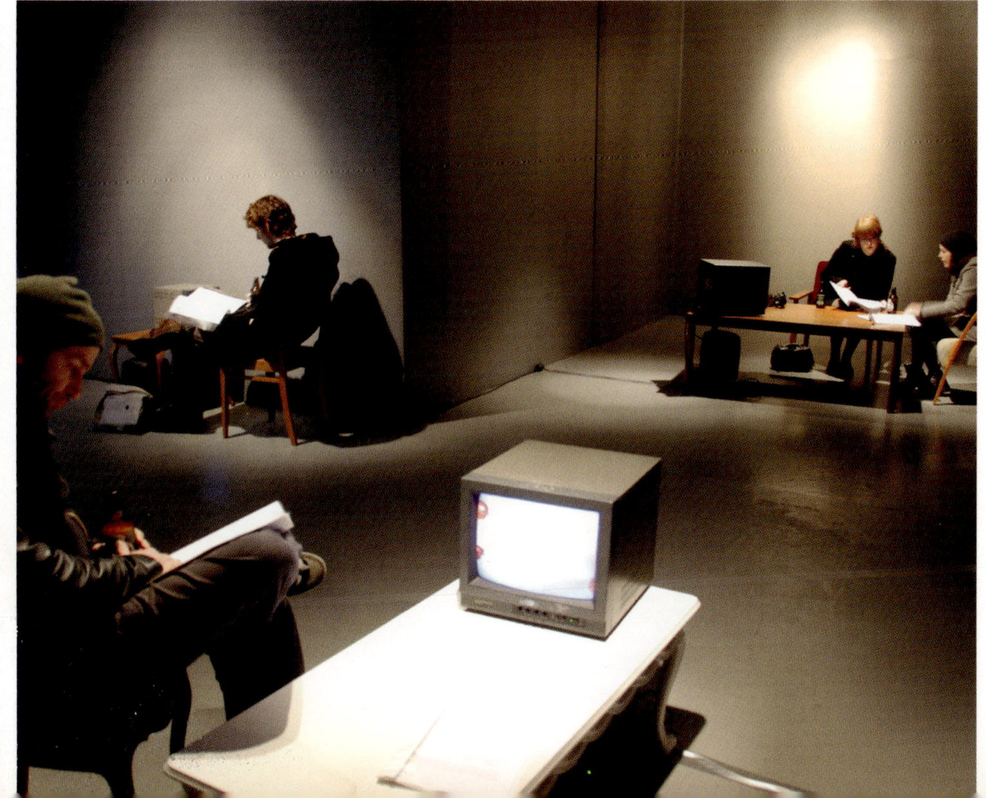

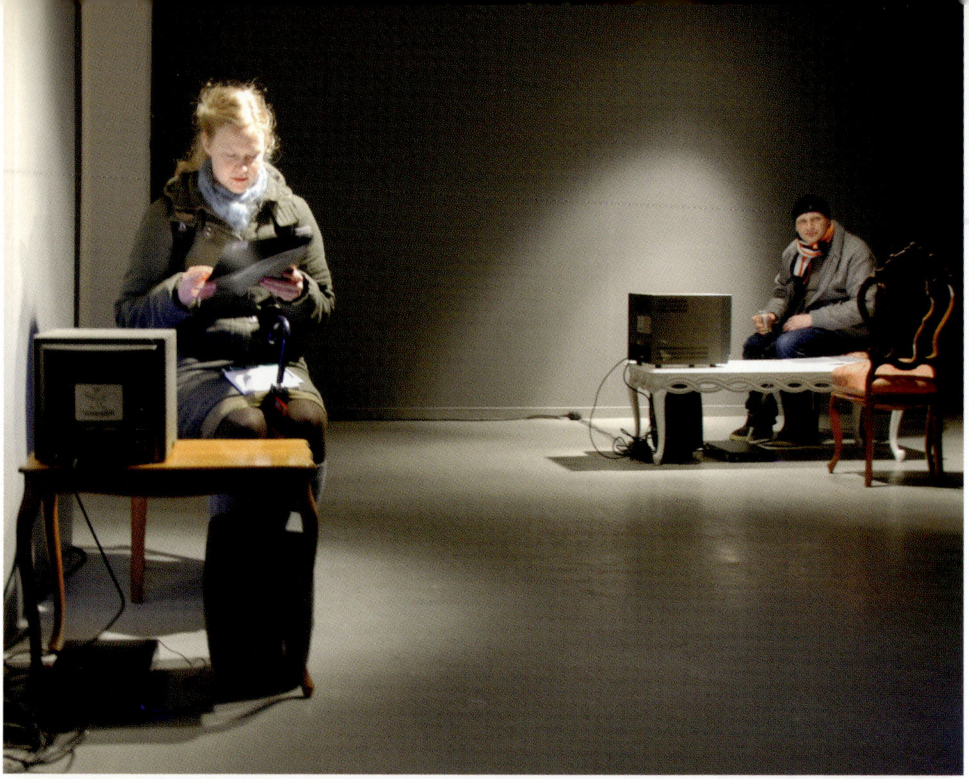

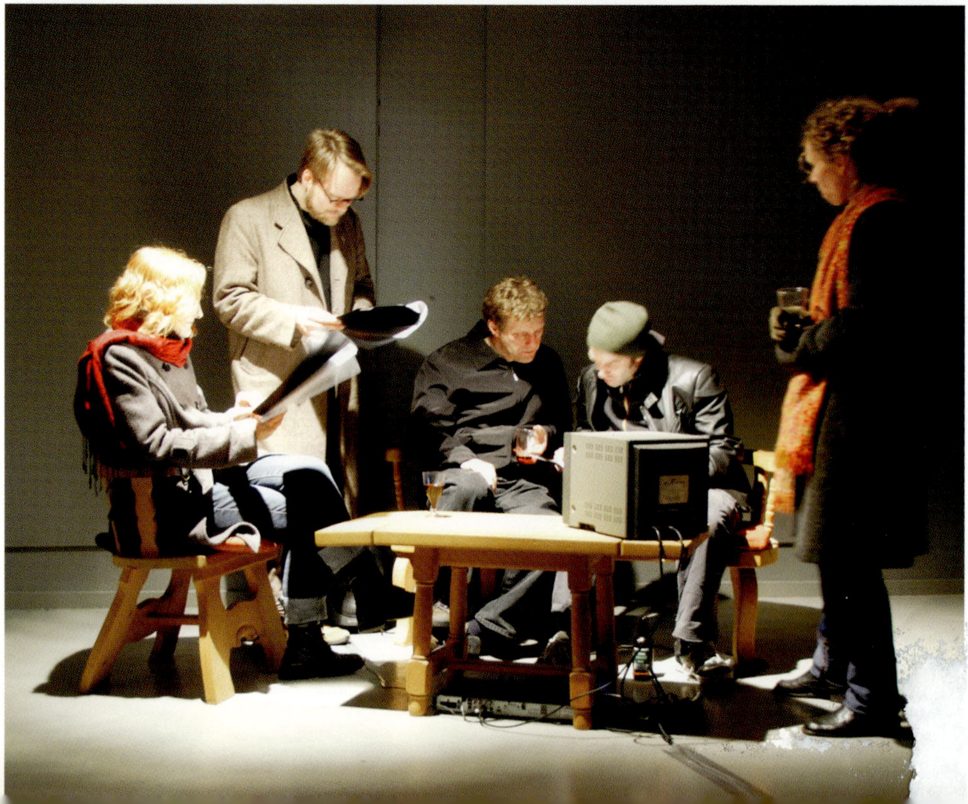

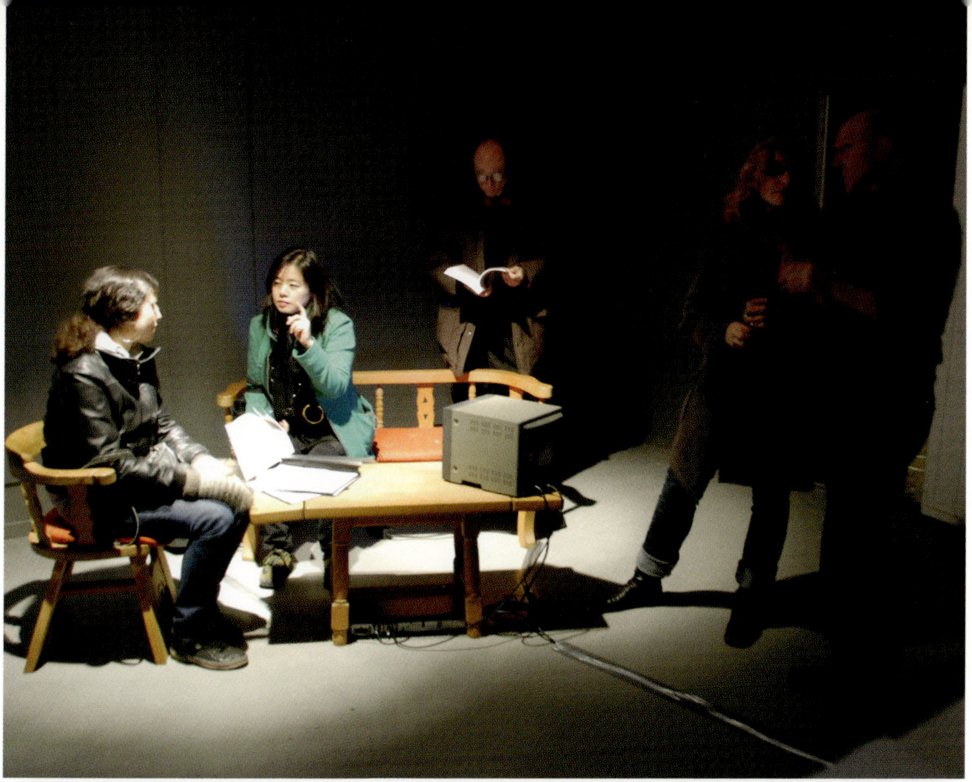

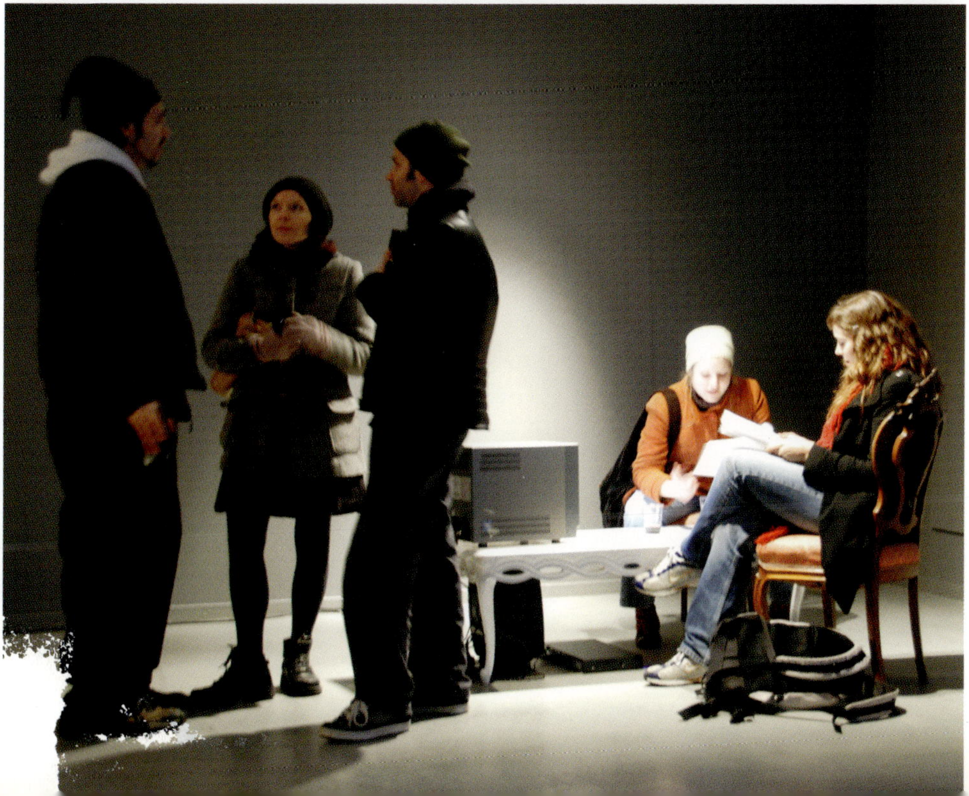

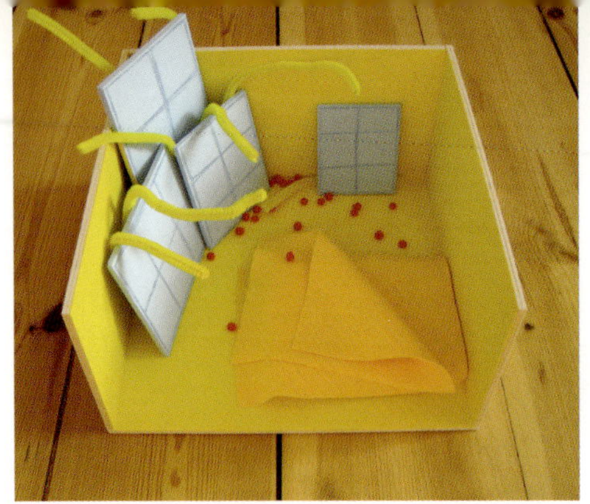

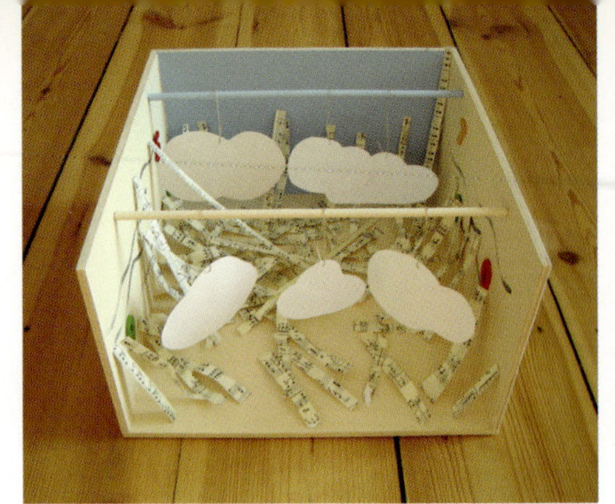

1

2

M: No … but it has a sensation to it … like a voice … but not really … it is spinning around me and giving me something … it gives me assurance … this light … I rest in this light … I rest in its yellow beaming glow … I see a window now … I recognize something … I am somewhere … somewhere I have been before … maybe when I was younger … not a kid anymore … but older … but before now … when I was younger … somewhere before … a yellow room … no … yes … blue outside … blue and yellow all around … I know I have seen this before … or I have felt this before … this being awake but sleeping too … something more brilliant than before … a condition … a sensation …

I watched out the window, at the tree in the front yard, the houses across the street, and the big white clouds in the blue sky, everything normal and still, ordinary and simple, and yet there was this sound, this voice, this music, carrying some secret message to my ears, changing how the houses looked, how the tree moved, how I saw things, my marbles, the wallpaper, even the sound of my mother doing the dishes… all of this became different, with this song and its brightness, its movement that touched a nerve in my chest and soared through to my legs, giving me a new feeling, a sensation that I can only describe as urgent.

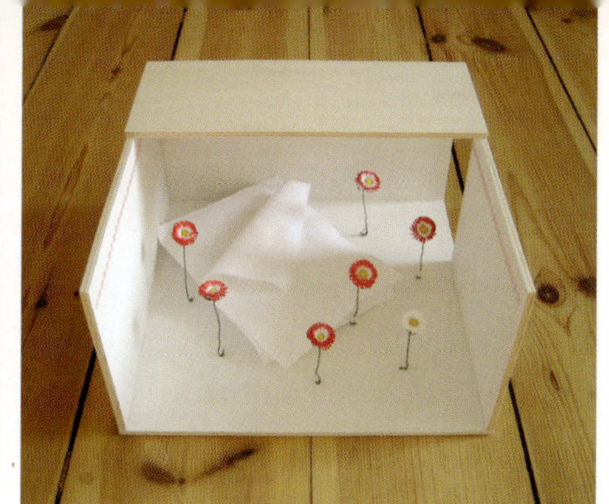 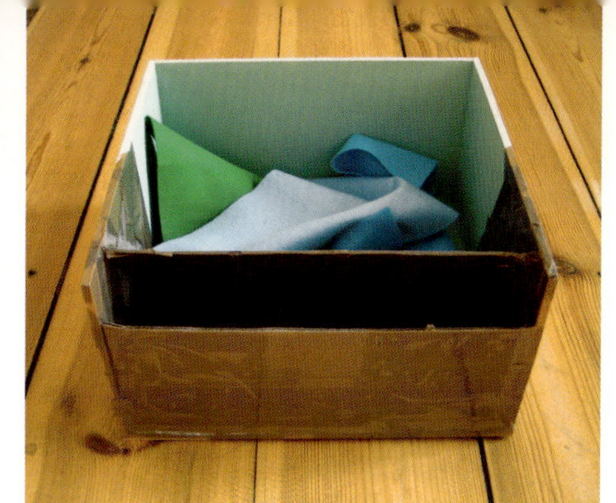

3

4

Q: So, you feel that the memory is connected to the music?

A: O yes, totally! I mean, the dream, the
flowers, the feeling, I am sure that
really all of this must have happened
in a period of a few minutes really,
probably the length of that song, because I knew that song, as my mother
always sang it when it was on the
radio, and it made her so happy, and
when she was happy I was happy too,
it was all connected, and I'm sure
that the dream happened because I
was hearing the song, while sleeping... it was there, and I'm sure it
made the dream possible, it created
that beautiful feeling inside.

You: Let's use the blue one.

Me: O yes!

You: I think if we put this one there, and
then pull this over like this... You
see, I think this will hold.

Me: O yes!

You: O but wait, maybe it could be better to have the green one under, that
way we can pretend we are in a forest, like with trees overhead.

Me: O yes! That's excellent.

You: Shhh... we have to be quiet, or we'll
wake up my parents.

Me: Ok.

You: Let's put the lamp just like this, underneath so we can see the dial.

Me: Ok.

CD: RADIO MEMORY / SCRIPT FOR A REHEARSAL

Working with four of the memories, I developed these into dramatic scripts – expanding the original narrative into an extended dialogue or monologue, accentuating the details and the scene while also allowing my own interpretation and personal relation to the memories to surface. The final works on the CD include readings by various friends and colleagues, and are composed as forms of radio drama.

Speakers: Fernanda Farah, Derek Holzer, Brandon LaBelle, Wolfgang Mayer, Chico Mello, Mark Robins, Steve Rowell and Annette Stahmer.

1. *The Bogus Man*

2. *Solsbury Hill*

3. *Morena de Angola*

4. *I am a Baby (in my Universe)*